The Lineman's and Cableman's Field Manual

Thomas M. Shoemaker, P.E., B.S.E.E.

*Life Senior Member, IEEE; Consulting Engineer;
Formerly: Manager, Distribution Department,
Iowa-Illionis Gas & Electric Company;
Member, Transmission and Distribution
Committee, Edison Electric Institute;
Captain, Signal Corps, U.S. Army*

James E. Mack, P.E., B.S.E.E., M.B.A.

*Manager, Electric Reliability,
MidAmerican Energy Company;
Senior Member, IEEE;
Member, N.S.P.E.*

Second Edition

New York   Chicago   San Francisco   Lisbon   London   Madrid
Mexico City   Milan   New Delhi   San Juan   Seoul
Singapore   Sydney   Toronto

The McGraw·Hill Companies

Cataloging-in-Publication Data is on file with the Library of Congress

1 2 3 4 5 6 7 8 9 0   DOC/DOC   0 1 5 4 3 2 1 0 9

ISBN 978-0-07-162121-2
MHID 0-07-162121-0

Sponsoring Editor
Stephen S. Chapman

Editing Supervisor
Stephen M. Smith

Production Supervisor
Pamela A. Pelton

Project Manager
Smita Rajan, International Typesetting and Composition

Copy Editor
Shivani Arora, International Typesetting and Composition

Proofreader
Priyanka Sinha, International Typesetting and Composition

Art Director, Cover
Jeff Weeks

Composition
International Typesetting and Composition

Printed and bound by RR Donnelley.

Contents

Preface

The Lineman's and Cableman's Field Manual is written to accompany *The Lineman's and Cableman's Handbook*. The *Field Manual* contains many of the same pertinent tables, charts, formulas, and safety rules that are found in the *Handbook*. The goal of the *Field Manual* is to provide the information that is needed on the job site in a format that can be quickly and easily accessed. This customizable tool provides specially designed notes pages for personal notes. It is written for the apprentice, the lineman, the cableman, the foreman, the supervisor, and other employees of electric line construction contractors and transmission and distribution departments of electric utility companies. The *Field Manual* contains definitions of electrical terms and diagrams of electric power systems, plus sections devoted to line conductors; cables, splices, and terminations; distribution voltage transformers; guying; lightning and surge protection; fuses; inspection and maintenance plans; tree trimming; rope, knots, splices, and gear; grounding; protective grounds; and safety equipment and rescue.

Safety is emphasized throughout this book. Occupational Safety and Health Administration (OSHA) Regulation (Standards—29 CFR) 1910.269, Electric Power Generation, Transmission, and Distribution, has been included due to its significance to the industry and for the safety of the individual. Of course, understanding the principles involved in any operation and knowing the reasons for doing things a given way are the best aids to safety. The opinion has become quite firmly established that a person is not a good lineman unless he does his work in accordance with established safety procedures and without injury to him or others. It is necessary for those engaged in electrical work to know the safety rules and the precautions applicable to their trades as specified in the *National Electrical Safety Code*, OSHA standards, and their employer's safety manuals and standards. Observing safety rules and procedures must be an inseparable part of their working habits.

This *Field Manual* places emphasis on the *National Electrical Safety Code*, OSHA standards, ANSI standards, and ASTM standards. Important requirements of all of these are discussed, but also they should be studied for detailed work procedures. Many applicable codes and standards are specified throughout the text to assist the reader.

The lineman and the cableman must become acquainted with the minimum construction requirements and maintenance and operating procedures in the various codes and standards to ensure the safety of the public and all workers. A copy of the *National Electrical Safety Code* (ANSI C2) can be secured for a fee from the Institute of Electrical and Electronics Engineers, Inc., 3 Park Avenue, 17th Floor, New York, NY 10016. All linemen and cablemen must know the information in the *National Electrical Safety Code* and adhere to the rules and procedures while performing their work assignments.

The *National Electrical Code*® details the rules and regulations for electrical installations, except those under the control of an electric utility. It excludes any indoor facility used and controlled exclusively by a utility for all phases from the generation through the distribution of electricity and for communication and metering, as well as outdoor facilities on a utility's own or a leased site or on public or private (by established rights) property.

The editors are well aware that one cannot become a competent lineman or cableman from a study of the pages of this book alone. However, diligent study along with daily practical experience and observation should give the apprentice an understanding of construction and maintenance procedures—and a regard for safety—that should make his progress and promotion rapid.

Thomas M. Shoemaker
James E. Mack

Acknowledgments

The editors are deeply indebted to several companies noted throughout the *Field Manual* and their representatives for contributions. The editors are also indebted to the Institute of Electrical and Electronics Engineers, Inc., publisher of the *National Electrical Safety Code;* the American National Standards Institute; *Electric Light and Power;* the Rural Electrification Administration; McGraw-Hill, publisher of the *Standard Handbook for Electrical Engineers,* and of *Electrical World; Transmission & Distribution;* and the Edison Electrical Institute, for permission to reprint various items from its literature.

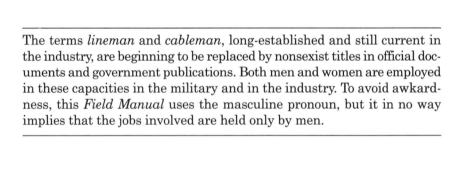

The terms *lineman* and *cableman*, long-established and still current in the industry, are beginning to be replaced by nonsexist titles in official documents and government publications. Both men and women are employed in these capacities in the military and in the industry. To avoid awkardness, this *Field Manual* uses the masculine pronoun, but it in no way implies that the jobs involved are held only by men.

The Electric System

Electrical Drawing Symbols

Diagrams of electric circuits show the manner in which electrical devices are connected. Because it would be impossible to make a pictorial drawing of each device shown in a diagram, the device is represented by a symbol.

A list of some of the most common electrical symbols used follows. It is noted that the symbol is a sort of simplified picture of the device represented.

The reader is encouraged to review *IEEE Standard 315/ANSI Y32.2/CSA Z99, Graphic Symbols for Electrical and Electronics Diagrams.*

Electrical device	Symbol	Electrical device	Symbol
Air circuit breaker		Cable termination	
Air circuit breaker drawout type, single pole		Capacitor	
three pole		Coil Electromagnetic actuator	
Air circuit breaker with magnetic-overload device		Asterisk represents device designation	

Electrical device	Symbol	Electrical device	Symbol
Air circuit breaker with thermal-overload device		Contact Make-before-break	
Ammeter		Normally closed	
Autotransformer		Normally open	

Battery		Contact (spring-return)
		  Pushbutton operated
		  Circuit closing (make)	

Battery cell		Circuit opening (break) Two circuit	

Contactor
  Electrically operated
  with series blowout coil
		Lightning arrester with Gap
		                    Valve
		                    Ground	

Contactor
  Electrically operated
  three pole with
  series blowout coils
  and auxiliary contacts—
  one closed, two open
		Magnetic blowout coil	

Current transformer		Network protector	
		Motor	

Current transformer
  with polarity
  marking
		Circuit breaker	C.B.

		Polarity markings
		  Negative
		  Positive	

Fuse		Potential transformer	

Gap
  horn
		Potential transformer
		  with
		  polarity markings	

Gap protective			
Generator		Rectifier	
Ground connection			

Electrical device	Symbol	Electrical device	Symbol

Inductance
  Fixed
		Resister
		  Fixed	

Adjustable		Tapped	
Lamp Red		Variable	
Green		With adjustable contact	

Switch
  Double throw
  Single throw
		Transformer
		1-phase with
		three windings	

		Transistor PNP-type	(E) (C) (B)

Transformer
Single phase
  two winding
		Electron tube
		  Triode	

Transformer connections
Three phase
  Delta
		Rectifier
		  two element	

Four wire grounded		Voltage regulator General	
Open		Induction	
		Single phase	
Three wire		Three phase	
Scott or T		Step	
Wye		Voltmeter	
Four wire grounded		Wattmeter	
Three wire ungrounded		Wires Connected	
Zigzag Grounded		Crossed	

Pictorial Representation
of an Electric System

For a pictorial overview of the different parts of an electric system, see
Figs. 1.1 through 1.4.

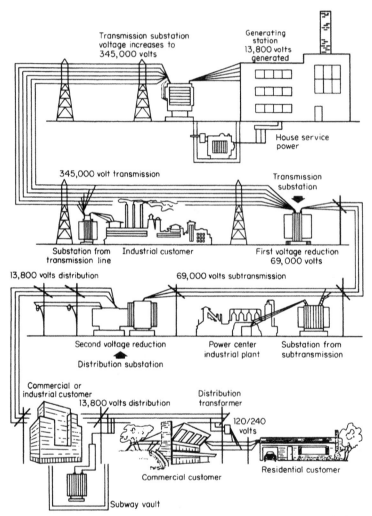

Figure 1.1   A simple model of the entire process of generation, transmission,
and distribution of electricity.

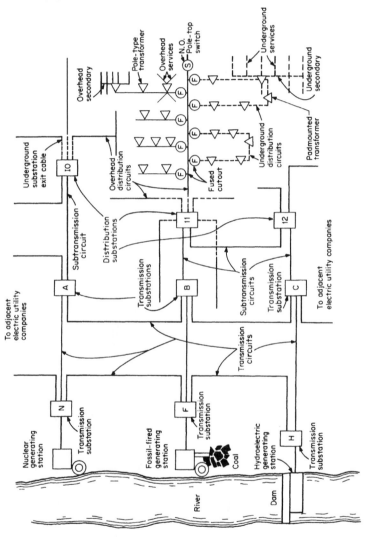

Figure 1.2 A single-line schematic of the entire process.

5

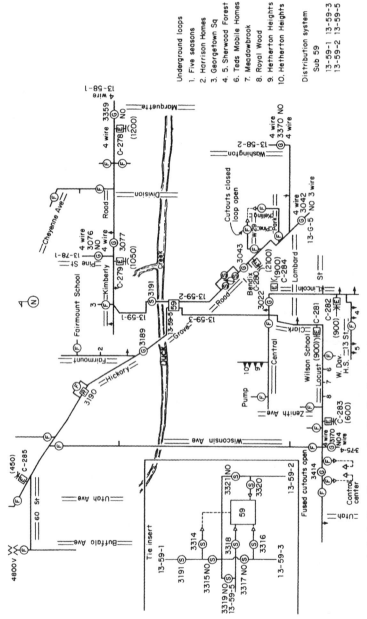

Figure 1.3 A single-line diagram of a distribution circuit.

6

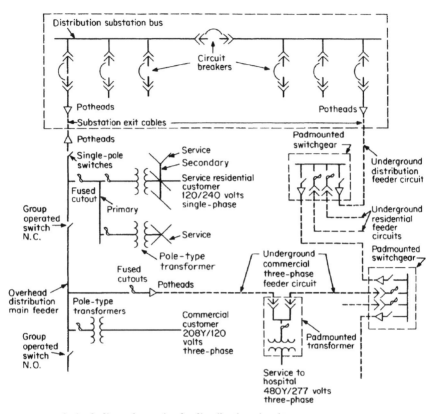

Figure 1.4 A single-line schematic of a distribution circuit.

Units of Measurement

Time

1 minute = 60 seconds

1 hour   = 60 minutes = 3,600 seconds

1 day   = 24 hours   = 1,440 minutes = 86,400 seconds

60-hertz AC electric system (60 cycles per second)

1 second   = 60 cycles

100 milliseconds = 6 cycles

250 milliseconds = 15 cycles

500 milliseconds = 30 cycles

Length

1 foot = 12 inches
1 yard = 3 feet   = 36 inches
1 mile = 1,760 yards = 5,280 feet = 63,360 inches

Decimal units

milli- 10^{-3}
centi- 10^{-2}
deci- 10^{-1}
kilo- 10^{3}
mega- 10^{6}

Metric equivalents

U.S. to metric

1 inch = 0.0254 meter	= 2.54 centimeters	= 25.4 millimeters
2 inches = 0.0508 meter	= 5.08 centimeters	= 50.8 millimeters
3 inches = 0.0762 meter	= 7.62 centimeters	= 76.2 millimeters
4 inches = 0.1016 meter	= 10.16 centimeters	= 101.6 millimeters
5 inches = 0.1270 meter	= 12.70 centimeters	= 127.0 millimeters
6 inches = 0.1524 meter	= 15.24 centimeters	= 152.4 millimeters
7 inches = 0.1778 meter	= 17.78 centimeters	= 177.8 millimeters
8 inches = 0.2032 meter	= 20.32 centimeters	= 203.2 millimeters
9 inches = 0.2286 meter	= 22.86 centimeters	= 228.6 millimeters
10 inches = 0.2540 meter	= 25.40 centimeters	= 254.0 millimeters
11 inches = 0.2794 meter	= 27.94 centimeters	= 279.4 millimeters
1 foot = 0.3048 meter	= 30.48 centimeters	= 304.8 millimeters
1 yard = 0.9144 meter	= 91.44 centimeters	= 914.4 millimeters
1 mile = 1609.35 meters =	1.60935 kilometers	

Metric to U.S.

1 millimeter = 0.03937 inch
1 centimeter = 0.39370 inch
1 meter   = 1.09361 yards = 3.28083 feet = 39.3700 inches
1 kilometer = 0.62137 mile

Temperature

Centigrade to Fahrenheit	Fahrenheit to Centigrade
$°C = (5/9)(°F-32)$	$°F = (9/5)°C + 32$

0°C = 32°F	0°F = −17.8°C
10°C = 50°F	10°F = −12.2°C
20°C = 68°F	20°F = −6.7°C
30°C = 86°F	30°F = −1.1°C
40°C = 104°F	32°F = 0°C
50°C = 122°F	40°F = 4.4°C
60°C = 140°F	50°F = 10.0°C
70°C = 158°F	60°F = 15.6°C
80°C = 176°F	70°F = 21.1°C
90°C = 194°F	80°F = 26.7°C
100°C = 212°F	90°F = 32.2°C
	100°F = 37.8°C

Definitions

- **Voltage (E)**   The electric force of work required to move current through an electric circuit, measured in units of volts.
- **Volt (V)**   The unit of measurement of voltage in an electric circuit, denoted by the symbol V.
- **Current (I)**   The flow of free electrons in one general direction, measured in units of amps.
- **Amp (A)**   The unit of measurement of electric current, denoted by the symbol A.
- **Direct-Current (dc)**   Current that flows continually in one direction.
- **Alternating-Current (ac)**   Current that flows in a circuit in a positive direction and then reverses itself to flow in a negative direction.
- **Resistance (R)**   The electrical "friction" that must be overcome through a device in order for current to flow when voltage is applied.
- **Ohm (Ω)**   The unit of measurement of resistance in an electric circuit, denoted by the symbol Ω.
- **Inductance (H)**   The property that causes the current to lag the voltage, measured in units of henries.
- **Capacitance (F)**   The property that causes the current to lead the voltage, measured in units of farads.

- **Energy (Wh)** The amount of electric work (real power) consumed or utilized in an hour. The unit of measurement is the watt hour.

- **Power (P)** The combination of electric current and voltage causing electricity to produce work. Power is composed of two components: real power and reactive power.

- **Volt-Ampere (VA)** The unit of both real and reactive power in an electric circuit.

- **Real Power** The resistive portion of a load found by taking the cosine (Θ) of the angle that the current and voltage are out of phase.

- **Watt (W)** The unit of real power in an electric circuit.

- **Reactive Power** The reactive portion of a load, found by taking the sine (Θ) of the angle that the current and voltage are out of phase.

- **Var (Q)** The unit of reactive power in an electric circuit.

- **Power Factor (pf)** The ratio of real power to reactive power.

- **Frequency (f)** The number of complete cycles made per second, measured in units of hertz.

- **Hertz (Hz)** Units of frequency (equal to 1 cycle per second).

- **Conductors** Materials that have many free electrons and are good transporters for the flow of electric current.

- **Insulators** Materials that have hardly any free electrons and inhibit or restrict the flow of electric current.

- **Series Resistive Circuit** All of the resistive devices are connected to each other so that the same current flows through all the devices.

- **Parallel Resistive Circuit** Each resistive device is connected across a voltage source. The current in the parallel path divides and only a portion of the current flows through each of the parallel paths.

Formulas and Calculations

In this section, the most common electrical formulas are given and their use is illustrated with a problem. In each case the formula is first stated by using the customary symbols, the same expression is then stated in words, a problem is given, and finally the substitutions are made for the symbols in the formula from which the answer is calculated. Only those formulas which the lineman is apt to have use for are given. The formulas are divided into three groups: dc circuits, ac circuits, and electrical apparatus.

Direct-current circuits

Ohm's law. The formula for Ohm's law is the most fundamental of all electrical formulas. It expresses the relationship that exists in an electric circuit containing resistance only between the current flowing in the resistance, the voltage impressed on the resistance, and the resistance of the circuit:

$$I = \frac{E}{R}$$

where I = current, amps
E = voltage, volts
R = resistance, ohms

Expressed in words, the formula states that current equals voltage divided by resistance, or:

$$Amperes = \frac{volts}{ohms}$$

Example: How much direct-current will flow through a resistance of 10 ohms if the pressure applied is 120 volts direct-current?

Solution: $I = \dfrac{E}{R} = \dfrac{120}{10} = 12$ amps

Ohm's law involves three quantities: current, voltage, and resistance. Therefore, when any two are known, the third one can be found. The procedure for solving for current has already been illustrated. To find the voltage required to circulate a given amount of current through a known resistance, Ohm's law is written as:

$$E = IR$$

In words, the formula says that the voltage is equal to the current multiplied by the resistance; thus:

$$Volts = amperes \times ohms$$

Example: How much direct-current (dc) voltage is required to circulate 5 amps direct current through a resistance of 15 ohms?

Solution: $E = IR = 5 \times 15 = 75$ volts

In like manner, if the voltage and current are known and the value of resistance is to be found, the formula is:

$$R = \frac{E}{I}$$

Expressed in words, the formula says that the resistance is equal to the voltage divided by the current; thus:

$$Ohms = \frac{volts}{amperes}$$

Example: What is the resistance of an electric circuit if 120 volts direct-current causes a current of 5 amps to flow through it?

Solution: $R = \dfrac{E}{I} = \dfrac{120}{5} = 24$ ohms

Series-resistive circuit

Example: The 12-volt battery has three lamps connected in series (Fig. 1.5). The resistance of each lamp is 8 ohms. How much current is flowing in the circuit?

Solution: In a series-resistive circuit, all values of resistance are added together, as defined by the formula:

$$R_{series} = R_1 + R_2 + \cdots + R_n = R_{total}$$

Therefore: $R = 8 + 8 + 8 = 24$ ohms

and because $I = E/R$
$$= 12/24$$
$$= 0.5 \text{ amp}$$

Parallel-resistive circuit

Example: The 12-volt battery has three lamps connected in parallel (Fig. 1.6). The resistance of each lamp is 8 ohms. How much current is flowing in the circuit?

Solution: In a parallel-resistive circuit, all values of resistance can be added together, as defined by the formula:

$$1/R_{parallel} = 1/R_1 + 1/R_2 + \cdots + 1/R_n$$

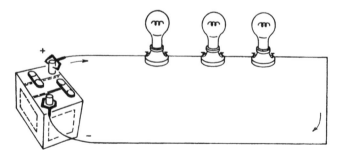

Figure 1.5 A series circuit. The same current flows through all the lamps.

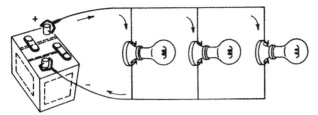

Figure 1.6 A parallel circuit. Each lamp is independent of the other lamps and draws its own current.

In our three-resistor example:

$$1/R_{\text{parallel}} = 1/R_1 + 1/R_2 + 1/R_3$$
$$1/R_{\text{parallel}} = (R_1R_2 + R_2R_3 + R_1R_3)/(R_1 \times R_2 \times R_3)$$
$$R_{\text{parallel}} = (R_1 \times R_2 \times R_3)/(R_1R_2 + R_2R_3 + R_1R_3)$$

Therefore: $R = (8 \times 8 \times 8)/[(8 \times 8) + (8 \times 8) + (8 \times 8)]$
$$= 512/192$$
$$= 2.67 \text{ ohms}$$

And because $I = E/R_{\text{parallel}}$
$$= 12/2.67$$
$$= 4.5 \text{ amps}$$

Power formula.   The expression for the power drawn by a dc circuit is:

$$P = EI$$

where E and I are the symbols for voltage in volts and current in amperes, respectively and P is the symbol for power in watts.

Expressed in words, the formula says that the power in watts drawn by a dc circuit is equal to the product of volts and amperes; thus:

$$Power = volts \times amperes$$

Example: How much power is taken by a 120-volt dc circuit when the current flowing is 8 amps direct-current?

Solution: $P = EI = 120 \times 8 = 960$ watts

This power formula also contains three quantities (terms): watts, volts, and amperes. Therefore, when any two of the three are known, the third one can be found. The procedure for finding the power when the voltage and current are given has already been illustrated. When the power and voltage are given and the current is to be found, the formula is:

$$I = \frac{P}{E}$$

In words, the formula states that the current equals the power divided by the voltage; thus:

$$Amperes = \frac{watts}{volts}$$

Example: How much direct-current would a 1000-watt load draw when connected to a 120-volt dc circuit?

Solution: $I = \dfrac{P}{E} = \dfrac{1000}{120} = 8.33$ amps

In like manner, when the power and current are known, the voltage can be found by writing the formula; thus:

$$E = \frac{P}{I}$$

Expressed in words, the formula says that the voltage is equal to the power divided by the current; thus:

$$Volts = \frac{watts}{amperes}$$

Example: What dc voltage would be required to deliver 660 watts with 6 amps direct-current flowing in the circuit?

Solution: $E = \dfrac{P}{I} = \dfrac{660}{6} = 110$ volts

Line loss or resistance loss. The formula for computing the power lost in a resistance when current flows through it is:

$$P = I^2R$$

where the symbols have the same meaning as in the foregoing formulas.

Expressed in words, the formula says that the power in watts lost in a resistance is equal to the square* of the current in amperes multiplied by the resistance in ohms; thus:

$$Watts = amperes\ squared \times ohms$$

Example: Compute the watts lost in a line having a resistance of 4 ohms when 8 amps direct-current is flowing in the line.

Solution: $P = I^2R = 8 \times 8 \times 4 = 256$ watts

---

*To *square* means to multiply by itself.

Energy (electrical work). The formula for computing the amount of energy consumed is defined as:

$$Energy = power \times time$$

Power is measured in units of watts and time is typically measured in units of hours, so the units for energy will be in wattshours.

Example: How much energy does a 60-watt load consume in a 10-hour time period?

The formula for energy is given by:

$$Energy = power \times time$$

Therefore:

$$Energy = 60 \text{ watts} \times 10 \text{ hours}$$
$$= 600 \text{ watthours}$$
$$= 0.6 \text{ kilowatthour}$$
$$= 0.6 \text{ kwh}$$

Alternating-current circuits

Ohm's law. Ohm's law is the same for resistance circuits when alternating voltage is applied as when direct voltage is applied, namely:

$$I = \frac{E}{R}$$

and

$$E = RI$$

and

$$R = \frac{E}{I}$$

In the preceding equations, E is the effective value of the alternating voltage and I the effective value of the alternating current. (See *Direct-current circuits* for examples.)

Ohm's law for other than resistance circuits. When alternating currents flow in circuits, these circuits might exhibit additional characteristics besides resistance. They might exhibit inductive reactance or capacitive reactance or both. The total opposition offered to the flow of current is then called *impedance* and is represented by the symbol Z. Ohm's law then becomes:

$$I = \frac{E}{Z}$$

and $$E = IZ$$

and $$Z = \frac{E}{I}$$

where I = current, amps
E = voltage, volts
Z = impedance, ohms

Example: Find the impedance of an alternating-current (ac) circuit if a 120-volt alternating voltage source causes 30 amps alternating current to flow.

Solution: $Z = \dfrac{E}{I} = \dfrac{120}{30}$ = 4 ohms impedance

Impedance.   The impedance of a series circuit is given by the expression:

$$Z = \sqrt{R^2 + (X_\mathrm{L} - X_\mathrm{c})^2}$$

where Z = impedance, ohms
R = resistance, ohms
X_L = inductive reatance, ohms
X_c = capacitive reactance, ohms

Example: If an ac-series circuit contains a resistance of 5 ohms, an inductive reactance of 10 ohms, and a capacitive reactance of 6 ohms, what is its impedance in ohms?

$$Z = \sqrt{R^2 + (X_\mathrm{L} - X_\mathrm{c})^2}$$
$$= \sqrt{5 \times 5 + (10 - 6)^2}$$
$$= \sqrt{25 + (4 \times 4)}$$
$$= \sqrt{41} = 6.4 \text{ ohms}$$

Note: The values of inductive and capacitive reactance of a circuit depend upon the frequency of the current, the size, spacing, and length of the conductors making up the circuit, etc. For distribution circuits and transmission lines, the values are located in the appropriate tables.

In the expression for Z, when neither inductive reactance nor capacitive reactance is present, Z reduces to the value R; thus:

$$Z = \sqrt{R^2 + (0 - 0)^2} = \sqrt{R^2} = R$$

Likewise, when only resistance and inductive reactance are present in the circuit, the expression for Z becomes:

$$Z = \sqrt{R^2 + (X_L - 0)^2} = \sqrt{R^2 + X_L^2}$$

Example: Compute the impedance of an ac circuit containing 3 ohms of resistance and 4 ohms of inductive reactance connected in series.

Solution: $Z = \sqrt{R^2 + X_L^2} = \sqrt{3^2 + 4^2}$

$\qquad = \sqrt{3 \times 3 + 4 \times 4} = \sqrt{9 + 16}$

$\qquad = \sqrt{25} = 5$ ohms impedance

Line loss or resistance loss.   The formula for computing the power lost in a resistance or line is:

$$P = I^2 R$$

where the symbols have the same meaning as above. Expressed in words, the formula says that the power in watts lost in a resistance is equal to the current in amperes squared multiplied by the resistance in ohms; thus:

$$Watts = amperes\ squared \times ohms$$

Example: Compute the power lost in a line having a resistance of 3 ohms if 20 amps is flowing in it.

Solution: $P = I^2 R = 20 \times 20 \times 3 = 1200$ watts

Power formula.   The power formula for single-phase ac circuits is:

$$P = E \times I \times pf$$

where $P =$ power, watts
$\qquad E =$ voltage, volts
$\qquad I =$ current, amps
$\qquad pf =$ power factor of circuit

Expressed in words, the formula says that the power (in watts) drawn by a single-phase ac circuit is equal to the product of volts, amperes, and power factor. If the power factor is unity, or 1, then the power equals the product of volts and amperes.

Example: What is the power drawn by a 20-amp purely resistive load, with a power factor of 1, operating at 120 volts?

Solution: $P = E \times I \times pf$
$= 120 \times 20 \times 1$
$= 2400$ watts
$= 2.4$ kilowatts
$= 2.4$ kW

Example: How much power is delivered to a single-phase ac circuit operating at 120 volts if the circuit draws 10 amps at 80 percent power factor?

Solution: $P = E \times I \times pf$
$= 12 \times 10 \times 0.80$
$= 960$ watts

The power formula given previously contains four quantities; P, E, I, and pf. Therefore, when any three are known, the fourth one can be determined. The case where the voltage, current, and power factor are known has already been illustrated. When the voltage, power, and power factor are known, the current can be computed from the expression:

$$I = \frac{P}{E \times pf}$$

Example: How many amperes are required to light an ordinary 60-watt, 120-volt incandescent lamp? (Assume that the power factor of a purely resistive load, such as the lamp, is 1.00.)

Solution: $I = \dfrac{P}{E \times pf}$
$= \dfrac{60}{120 \times 1}$
$= 0.5$ amp

Example: How much current is drawn by a 10-kW load from a 220-volt ac circuit if the power factor is 0.80?

Solution: $I = \dfrac{P}{E \times pf}$
$= \dfrac{10,000}{200 \times 0.80} = 56.8$ amps

In similar manner, the power, voltage, and current are known, and the power factor can be computed from the following formula:

$$pf = \frac{P}{E \times I}$$

Example: What is the power factor of a 4-kW load operating at 230 volts if the current drawn is 20 amps?

Solution: $pf = \dfrac{P}{E \times I} = \dfrac{4000}{230 \times 20}$

$= 0.87 = 87$ percent power factor

Three-phase.   The power formula for a three-phase ac circuit is:

$$P = \sqrt{3} \times E \times I \times pf$$

where E is the voltage between phase wires and the other symbols have the same meaning as for single-phase. The additional quantity in the three-phase formula is the factor $\sqrt{3}$, expressed "square root of three," the value of which is:

$$\sqrt{3} = 1.73$$

The voltage between phase wires is equal to $\sqrt{3}$ (or 1.73), multiplied by the phase-to-neutral voltage of the three-phase four-wire systems.

Example: How much power in watts is drawn by a three-phase load at 230 volts if the current is 10 amps and the power factor is 80 percent?

Solution: $P = \sqrt{3} \times E \times I \times pf$

$= \sqrt{3} \times 230 \times 10 \times 0.80$

$= 3183$ watts

The three-phase power formula also contains four quantities, as in the single-phase formula. Therefore, if any three are known, the fourth one can be determined. The case where the voltage, current, and power factor are known was illustrated previously. When the power, voltage, and power factor are known, the current in the three-phase line can be computed from the formula:

$$I = \frac{P}{\sqrt{3}E \times pf}$$

Example: If a 15-kW three-phase load operates at 2300 volts and has a power factor of 1, how much current flows in each line of the three-phase circuit?

Solution: $I = \dfrac{P}{\sqrt{3}E \times pf}$

$= \dfrac{15,000}{\sqrt{3} \times 2300 \times 1.0} = 3.77$ amps

If the only quantity not known is the power factor, this can be found by using the following formula:

$$pf = \frac{P}{\sqrt{3} \times E \times I}$$

Example: Find the power factor of a 100-kW three-phase load operating at 2300 volts if the line current is 40 amps.

Solution: $pf = \dfrac{P}{\sqrt{3} \times E \times I}$

$$= \frac{100,000}{\sqrt{3} \times 2300 \times 40}$$

$$= 0.628 = 62.8 \text{ percent power factor}$$

Volt-amperes. Often loads are given in volt-amperes instead of watts or in kVA instead of kW. The relationships then become:

$$VA = E \times I \qquad \text{for single-phase}$$
and
$$VA = \sqrt{3}E \times I \quad \text{for three-phase}$$

In using these quantities, it is not necessary to know the power factor of the load.

Power factor correction

To determine the kVARs required to improve the power factor for a known load, the following steps are followed:

Step 1: Determine the existing kVA load:

$$kVA_{existing} = \frac{kW}{pf_{existing}}$$

Step 2: Determine the existing reactive power load, expressed in kVARs:

$$kVAR_{existing} = \sqrt{kVA_{existing}^2 - kW^2}$$

Step 3: Determine the corrected kVA load:

$$kVA_{corrected} = \frac{kW}{pf_{corrected}}$$

Step 4: Determine the corrected reactive power load, expressed in kVARs:

$$kVAR_{corrected} = \sqrt{kVA_{corrected}^2 - kW^2}$$

Step 5: The required corrective reactive power is equal to the existing reactive power load minus the corrected reactive power load, expressed in kVARs:

$$kVAR_{required} = kVAR_{existing} - kVAR_{corrected}$$

Example: For a 40-kW load at 80% power factor, how much $kVAR$ is needed to correct the power factor to 90%?

The first step is to determine the kilovolt load in the existing condition:

$$kVA_{existing} = kW/pf_{existing}$$
$$= 40/0.8$$
$$= 50 \text{ kVA}$$

Step 2 is to determine the reactive power load in the existing condition:

$$kVAR_{existing} = \sqrt{kVA^2 - kW^2}$$
$$= \sqrt{50^2 - 40^2}$$
$$= 30 \text{ kVARs}$$

Having found the existing condition reactive power load, determine what the corrected final values for the total and reactive power will be:

$$\textbf{Step 3: } kVA_{corrected} = kW/pf_{corrected}$$
$$= 40/0.9$$
$$= 44.4 \text{ kVA}$$

$$\textbf{Step 4: } kVAR_{corrected} = \sqrt{kVA^2 - kW^2}$$
$$= \sqrt{44.4^2 - 40^2}$$
$$= 19.37 \text{ kVARs}$$

To determine the amount of reactive power needed on the system to correct the power factor from 80% to 90%, the final step of the problem is to subtract the corrected reactive power value desired from the existing value of reactive power:

$$\textbf{Step 5: } kVAR_{required} = kVAR_{existing} - kVAR_{corrected}$$
$$= 30 - 19.37$$
$$= 10.63 \text{ kVARs}$$

Line loss or resistance loss (three-phase). The power lost in a three-phase line is given by the expression:

$$P = 3 \times I^2 R$$

where I = current in each line wire and R = resistance of each line wire. The factor 3 is present so that the loss in a line wire will be taken three times to account for the loss in all three wires.

Example: If a three-phase line carries a current of 12 amps and each line wire has a resistance of 3 ohms, how much power is lost in the resistance of the line?

Solution: $P = 3 \times I^2R = 3 \times 12 \times 12 \times 3 = 1296$ watts
$$= 1.296 \text{ kilowatts}$$

Electrical Apparatus

Motors

Direct-current motor. The two quantities that are usually desired in an electric motor are its output in horsepower (hp) and its input current rating at full load. The following expression for current is used when the motor is a direct-current motor:

$$I = \frac{hp \times 746}{E \times eff}$$

The formula says that the full-load current is obtained by multiplying the horsepower by 746 and dividing the result by the product of voltage and percent efficiency. (The number of watts in 1 hp is 746. A kilowatt equals $1^{1}/_{3}$ horsepower.)

Example: How much current will a 5-hp 230-volt dc motor draw at full load if its efficiency at full load is 90 percent?

Solution: $I = \dfrac{hp \times 746}{E \times eff} = \dfrac{5 \times 746}{230 \times 0.90} = 18.0$ amps

Alternating-current motor, single-phase. If the motor is a single-phase ac motor, the expression for current is almost the same, except that it must allow for power factor. The formula for full-load current is:

$$I = \frac{hp \times 746}{E \times eff \times pf}$$

Example: How many amperes will a $^{1}/_{4}$-hp motor take at full load if the motor is rated 110 volts and has a full-load efficiency of 85 percent and a full-load power factor of 80 percent?

Solution: $I = \dfrac{hp \times 746}{E \times eff \times pf} = \dfrac{0.25 \times 746}{110 \times 0.85 \times 0.80} = 2.5$ amps

Alternating-current motor, three-phase. The formula for full-load current of a three-phase motor is the same as for a single-phase motor, except that it has the factor $\sqrt{3}$ in it; thus:

$$I = \frac{hp \times 746}{\sqrt{3} \times E \times eff \times pf}$$

Example: Calculate the full-load current rating of a 5-hp three-phase motor operating at 220 volts and having an efficiency of 80 percent and a power factor of 85 percent.

Solution: $I = \dfrac{hp \times 746}{\sqrt{3} \times E \times eff \times pf} = \dfrac{5 \times 746}{\sqrt{3} \times 220 \times 0.80 \times 0.85}$

= 14.4 amps

Direct-current motor.   To find the horsepower rating of a dc motor if its voltage and current rating are known, the same quantities are used as for current except the formula is rearranged thus:

$$hp = \frac{E \times I \times eff}{746}$$

Example: How many horsepower can a 220-volt dc motor deliver if it draws 15 amps and has an efficiency of 90 percent?

Solution: $hp = \dfrac{E \times I \times eff}{746} = \dfrac{220 \times 15 \times 0.90}{746} = 4.0 \text{ hp}$

Alternating-current motor, single-phase.   The horsepower formula for a single-phase ac motor is

$$hp = \frac{E \times I \times pf \times eff}{746}$$

Example: What is the horsepower rating of a single-phase motor operating at 480 volts and drawing 25 amps at a power factor of 88 percent if it has a full-load efficiency of 90 percent?

Solution: $hp = \dfrac{E \times I \times pf \times eff}{746}$

$= \dfrac{480 \times 25 \times 0.88 \times 0.90}{746}$

= 12.74 hp

Alternating-current motor, three-phase.   The horsepower formula for a three-phase ac motor is

$$hp = \frac{\sqrt{3} \times E \times I \times pf \times eff}{746}$$

Example: What horsepower does a three-phase 240-volt motor deliver if it draws 10 amps, has a power factor of 80 percent, and has an efficiency of 85 percent?

Solution:
$$hp = \frac{\sqrt{3} \times E \times I \times pf \times eff}{746}$$
$$= \frac{\sqrt{3} \times 240 \times 10 \times 0.80 \times 0.85}{746} = 3.8 \text{ hp}$$

Alternating-current generator

Frequency. The frequency of the voltage generated by an ac generator depends on the number of poles in its field and the speed at which it rotates:

$$f = \frac{p \times rpm}{120}$$

where f = frequency, Hz
p = number of poles in field
rpm = revolutions of field per minute

Example: Compute the frequency of the voltage generated by an alternator having two poles and rotating at 3600 rpm.

Solution: $f = \dfrac{p \times rpm}{120} = \dfrac{2 \times 3600}{120} = 60 \text{ Hz}$

Speed. To determine the speed at which an alternator should be driven to generate a given frequency, the following expression is used:

$$rpm = \frac{f \times 120}{p}$$

Example: At what speed must a four-pole alternator be driven to generate 60 Hz?

Solution: $rpm = \dfrac{f \times 120}{p} = \dfrac{60 \times 120}{4} = 1800 \text{ rpm}$

Number of poles. If the frequency and speed of an alternator are known, the number of poles in its field can be calculated by use of the following formula:

$$p = \frac{f \times 120}{rpm}$$

Example: How many poles does an alternator have if it generates 60 Hz at 1200 rpm?

Solution: $p = \dfrac{f \times 120}{rpm} = \dfrac{60 \times 120}{1200} = 6 \text{ poles}$

Transformer (single-phase)

Primary current. The full-load primary current can be readily calculated if the kVA rating of the transformer and the primary voltage are known:

$$I_P = \frac{kVA \times 1000}{E_P}$$

where E_P = rated primary voltage and I_P = rated primary current.

Example: Find the rated full-load primary current of a 10-kVA 2300-volt distribution transformer.

Solution: $I_P = \dfrac{kVA \times 1000}{E_P} = \dfrac{10 \times 1000}{2300} = 4.3 \text{ amps}$

Secondary current. The expression for secondary current is similar to that for primary current. Thus:

$$I_s = \frac{kVA \times 1000}{E_s}$$

where I_s = rated secondary current and E_s = rated secondary voltage.

Example: A 3-kVA distribution transformer is rated 2300 volts primary and 110 volts secondary. What is its full-load secondary current?

Solution: $I_s = \dfrac{kVA \times 1000}{E_s}$

$= \dfrac{3 \times 1000}{110} = 27 \text{ amps secondary}$

Summary of Electrical Formulas

For a summary of circuit calculations, see Table 1.1.

TABLE 1.1 Summary of Circuit Calculations

To find	Direct-current	Alternating-current Single-phase	Alternating-current Three-phase
Amperes when kilovolt-amperes and voltage are known	$\dfrac{Kilovolt\text{-}amperes \times 1000}{Volts}$	$\dfrac{Kilovolt\text{-}amperes \times 1000}{Volts}$	$\dfrac{Kilovolt\text{-}amperes \times 1000}{1.73 \times volts}$
Kilovolt-amperes when amperes and voltage are known	$\dfrac{Amperes \times volts}{1000}$	$\dfrac{Amperes \times volts}{1000}$	$\dfrac{1.73 \times volts \times amperes}{1000}$
Amperes when kilowatts, volts, and power factors are known	$\dfrac{Kilowatts \times 1000}{Volts}$	$\dfrac{Kilowatts \times 1000}{Volts \times power\ factor}$	$\dfrac{Kilowatts \times 1000}{1.73 \times volts \times power\ factor}$
Kilowatts when amperes, volts, and power factor are known	$\dfrac{Amperes \times volts}{1000}$	$\dfrac{Amperes \times volts \times power\ factor}{1000}$	$\dfrac{1.73 \times amperes \times volts \times power\ factor}{1000}$

Current

1. *Direct-current.* To find the current when the voltage and resistance are given:

$$I = \frac{E}{R}$$

2. *Alternating-current.* To find the current when the voltage and resistance are given:

$$I = \frac{E}{R}$$

3. *Alternating-current.* To find the current when the voltage and impedance are given:

$$I = \frac{E}{Z}$$

4. *Direct-current.* To find the current when the voltage and power are given:

$$I = \frac{P}{E}$$

5. *Alternating-current, single-phase.* To find the current when the voltage, power, and power factor are given:

$$I = \frac{P}{E \times pf}$$

6. *Alternating-current, three-phase.* To fine the current when the voltage, power, and power factor are given:

$$I = \frac{P}{\sqrt{3} \times pf}$$

7. *Alternating-current, single-phase.* To find the current when the volt-amperes and volts are given:

$$I = \frac{VA}{E}$$

and
$$I = \frac{kVA \times 1000}{E}$$

8. *Alternating-current, three-phase.* To find the current when the volt-amperes and volts are given:

$$I = \frac{VA}{\sqrt{3}E}$$

and
$$I = \frac{kVA \times 1000}{\sqrt{3}E}$$

Voltage

9. *Direct-current.* To find the voltage when the current and resistance are given:

$$E = IR$$

10. *Alternating-current.* To find the voltage when the current and resistance are given:

$$E = IR$$

11. *Alternating-current.* To find the voltage when the current and impedance are given:

$$E = IZ$$

Resistance

12. *Direct-current.* To find the resistance when the voltage and current are given:

$$R = \frac{E}{I}$$

13. *Alternating-current.* To find the resistance when the voltage and current are given:

$$R = \frac{E}{I}$$

Power loss

14. *Direct-current.* To find the power loss when the current and resistance are given:

$$P = I^2 R$$

15. *Alternating-current, single-phase.* To find the power loss when the current and resistance are given:

$$P = I^2 R$$

16. *Alternating-current, three-phase.* To find the power loss when the current and resistance are given:

$$P = 3 \times I^2 R$$

Impedance

17. *Alternating-current, single-phase.* To find the impedance when the current and voltage are given:

$$Z = \frac{E}{I}$$

18. *Alternating-current, single-phase.* To find the impedance when the resistance and inductive reactance are given:

$$Z = \sqrt{R^2 + X_L^2}$$

19. *Alternating-current, single-phase.* To find the impedance when the resistance, inductive reactance, and capacitive reactance are given:

$$Z = \sqrt{R^2 + (X_L - X_c)^2}$$

Power

20. *Direct-current.* To find the power when the voltage and current are given:

$$P = EI$$

21. *Alternating-current, single-phase.* To find the power when the voltage, current, and power factor are given:

$$P = E \times I \times pf$$

22. *Alternating-current, three-phase.* To find the power when the voltage, current, and power factor are given:

$$P = \sqrt{3}E \times I \times pf$$

Volt-amperes

23. *Alternating-current, single-phase.* To find the volt-amperes when the voltage and current are given:

$$VA = E \times I$$

and
$$kVA = \frac{E \times I}{1000}$$

24. *Alternating-current, three-phase.* To find the volt-amperes when the voltage and current are given:

$$VA = \sqrt{3}E \times I$$

and
$$kVA = \frac{\sqrt{3}E \times I}{1000}$$

Power factor

25. *Alternating-current, single-phase.* To find the power factor when the power, volts, and amperes are given:

$$pf = \frac{P}{E \times I}$$

26. *Alternating-current, three-phase.* To find the power factor when the power, volts, and amperes are given:

$$pf = \frac{P}{\sqrt{3} \times E \times I}$$

Motor

27. *Direct-current motor.* To find the current when the horsepower of the motor, the voltage, and the efficiency are given:

$$I = \frac{hp \times 746}{E \times eff}$$

28. *Alternating-current motor, single-phase.* To find the current when the horsepower of the motor, the voltage, the efficiency, and the power factor are given:

$$I = \frac{hp \times 746}{E \times pf \times eff}$$

NOTES

NOTES

2

Line Conductors

Types

- Copper
- Aluminum
- Steel

Definitions

Solid conductor   A single conductor of solid circular section (Fig. 2.1).

Stranded conductor   A group of wires made into a single conductor (Fig. 2.2).

Aluminum conductor steel reinforced (ACSR)   (Fig. 2.3 and Tables 2.1 and 2.2)

All-aluminum conductor (AAC)   (Tables 2.3 and 2.4)

Aluminum conductor steel supported (ACSS)   (Tables 2.5 and 2.6)

All-aluminum alloy conductor (AAAC)   (Table 2.7)

Aluminum conductor alloy reinforced (ACAR)   (Table 2.8)

Aluminum conductor composite reinforced (ACCR)

Aluminum conductor composite core (ACCC)

Trapezoidal wire (TW)

Vibration-resistant (VR) conductor

Self-damping (SD)

Copperweld steel wire   A coating of copper is welded to the steel wire (Fig. 2.4).

Alumoweld steel wire   A coating of aluminum is welded to the steel wire (Fig. 2.5).

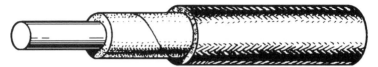

Figure 2.1 Solid conductor with two layer covering.

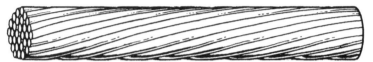

Figure 2.2 Stranded conductor.

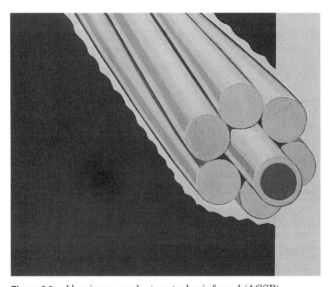

Figure 2.3 Aluminum conductor steel reinforced (ACSR).

Figure 2.4 Copperweld steel wire.

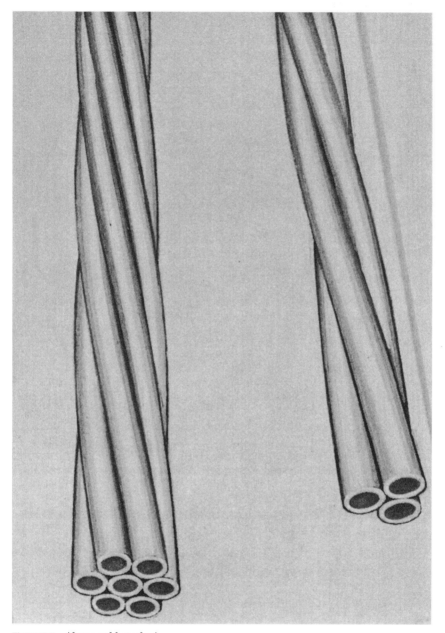

Figure 2.5   Alumoweld steel wire.

TABLE 2.1 Aluminum Conductor Steel Reinforced—Physical Properties

Code word	Size AWG or kcmil	Stranding number & diameter (in) Aluminum	Stranding number & diameter (in) Steel	Nominal diameter (in) Complete conductor	Nominal diameter (in) Steel core	Rated strength (lb)	Weight (lb/1000 ft) Total	Weight (lb/1000 ft) Aluminum	Weight (lb/1000 ft) Steel	Percent of total weight Aluminum	Percent of total weight Steel
TURKEY	6	6 × 0.0661	1 × 0.0661	0.198	0.0661	1190	36.0	24.4	11.6	67.8	32.2
SWAN	4	6 × 0.0834	1 × 0.0834	0.250	0.0834	1860	57.4	39.0	18.4	67.9	32.1
SWANATE	4	7 × 0.0772	1 × 0.1029	0.257	0.1029	2360	67.0	39.0	28.0	58.2	41.8
SPARROW	2	6 × 0.1052	1 × 0.1052	0.316	0.1052	2850	91.2	61.9	29.3	67.9	32.1
SPARATE	2	7 × 0.0974	1 × 0.1299	0.325	0.1299	3640	106.6	61.9	44.7	58.1	41.9
ROBIN	1	6 × 0.1181	1 × 0.1181	0.354	0.1181	3550	115.0	78.1	36.9	67.9	32.1
RAVEN	1/0	6 × 0.1327	1 × 0.1327	0.398	0.1327	4380	145.2	98.6	46.6	67.9	32.1
QUAIL	2/0	6 × 0.1489	1 × 0.1489	0.447	0.1489	5300	182.8	124.1	58.7	67.9	32.1
PIGEON	3/0	6 × 0.1672	1 × 0.1672	0.502	0.1672	6620	230.5	156.4	74.1	67.9	32.1
PENGUIN	4/0	6 × 0.1878	1 × 0.1878	0.563	0.1878	8350	290.8	197.4	93.4	67.9	32.1
WAXWING	266.8	18 × 0.1217	1 × 0.1217	0.509	0.1217	6880	289.1	249.9	39.2	86.4	13.6
PARTRIDGE	266.8	26 × 0.1013	7 × 0.0788	0.642	0.236	11300	366.9	251.3	115.6	68.5	31.5
MERLIN	336.4	18 × 0.1367	1 × 0.1367	0.684	0.1367	8700	364.8	315.3	49.5	86.4	13.6
LINNET	336.4	26 × 0.1137	7 × 0.0884	0.720	0.265	14100	462.0	316.5	145.5	68.5	31.5
ORIOLE	336.4	30 × 0.1059	7 × 0.1059	0.741	0.318	17300	526.4	317.7	208.7	60.4	39.6
CHICKADEE	397.5	18 × 0.1486	1 × 0.1486	0.743	0.1486	9900	431.0	372.5	58.5	86.4	13.6
IBIS	397.5	26 × 0.1236	7 × 0.0961	0.783	0.288	16300	546.0	374.1	171.9	68.5	31.5
LARK	397.5	30 × 0.1151	7 × 0.1151	0.806	0.345	20300	621.8	375.2	246.6	60.4	39.6
PELICAN	477	18 × 0.1628	1 × 0.1628	0.814	0.1628	11800	517.3	447.1	70.2	86.4	13.6
FLICKER	477	24 × 0.1410	7 × 0.0940	0.846	0.282	17200	613.9	449.4	164.5	73.2	26.8
HAWK	477	26 × 0.1354	7 × 0.1053	0.858	0.316	19500	655.3	448.9	206.4	68.5	31.5
HEN	477	30 × 0.1261	7 × 0.1261	0.883	0.378	23800	746.4	450.4	296.0	60.4	39.6
OSPREY	556.5	18 × 0.1758	1 × 0.1758	0.879	0.1758	13700	603.3	521.4	81.9	86.4	13.6
PARAKEET	556.5	24 × 0.1523	7 × 0.1015	0.914	0.304	19800	716.1	524.3	191.8	73.2	26.8
DOVE	556.5	26 × 0.1463	7 × 0.1138	0.927	0.341	22600	765.2	524.2	241.0	68.5	31.5
EAGLE	556.5	30 × 0.1362	7 × 0.1362	0.953	0.409	27800	870.7	525.4	345.3	60.4	39.6

PEACOCK	605	24 × 0.1588	7 × 0.1059	0.953	0.318	21600	778.8	570.1	208.7	73.2	26.8
SWIFT	636	36 × 0.1329	1 × 0.1329	0.930	0.1329	13800	642.8	596.0	46.8	92.7	7.3
KINGBIRD	636	18 × 0.1880	1 × 0.1880	0.940	0.1880	15700	689.9	596.3	93.6	86.4	13.6
ROOK	636	24 × 0.1628	7 × 0.1085	0.977	0.326	22600	818.2	599.1	219.1	73.2	26.8
GROSBEAK	636	26 × 0.1564	7 × 0.1216	0.990	0.365	25200	874.2	599.0	275.2	68.5	31.5
EGRET	636	30 × 0.1456	19 × 0.0874	1.019	0.437	31500	987.2	600.5	386.7	60.8	39.2
FLAMINGO	666.6	24 × 0.1667	7 × 0.111	1.000	0.333	23700	857.9	628.2	229.7	73.2	26.8
STARLING	715.5	26 × 0.1659	7 × 0.1290	1.051	0.387	28400	983.7	674.0	309.7	68.5	31.5
REDWING	715.5	30 × 0.1544	19 × 0.0926	1.081	0.463	34600	1109.3	675.3	434.0	60.8	39.2
COOT	795	36 × 0.1486	1 × 0.1486	1.040	0.1486	16800	803.6	745.1	58.5	92.7	7.3
TERN	795	45 × 0.1329	7 × 0.0886	1.063	0.266	22100	895	749	146	83.7	16.3
CUCKOO	795	24 × 0.1820	7 × 0.1213	1.092	0.364	27900	1023	749	274.0	73.2	26.8
CONDOR	795	54 × 0.1213	7 × 0.1213	1.092	0.364	28200	1022	748	274.0	73.2	26.8
DRAKE	795	26 × 0.1749	7 × 0.1360	1.108	0.408	31500	1093	749	344	68.5	31.5
MALLARD	795	30 × 0.1628	19 × 0.0977	1.140	0.489	38400	1233.9	750.7	483.2	60.8	39.2
RUDDY	900	45 × 0.1414	7 × 0.0943	1.131	0.283	24400	1013	848	165	83.7	16.3
CANARY	900	54 × 0.1291	7 × 0.1291	1.162	0.387	31900	1158	848	310	73.2	26.8
CORNCRAKE	954	20 × 0.2184	7 × 0.0971	1.165	0.291	25600	1074	899	175	83.7	16.3
REDBIRD	954	24 × 0.1994	7 × 0.1329	1.196	0.399	33500	1228	899	329	73.2	26.8
TOWHEE	954	48 × 0.1410	7 × 0.1097	1.175	0.329	28500	1123	899	224	80.1	19.9
RAIL	954	45 × 0.1456	7 × 0.0971	1.165	0.291	25900	1075	899	176	83.7	16.3
CARDINAL	954	54 × 0.1329	7 × 0.1329	1.196	0.399	33800	1227.1	898.4	328.7	73.2	26.8
ORTOLAN	1033.5	45 × 0.1515	7 × 0.1010	1.212	0.303	27700	1163	973	190	83.7	16.3
CURLEW	1033.5	54 × 0.1383	7 × 0.1383	1.245	0.415	36600	1329	973	356	73.2	26.8
BLUEJAY	1113.0	45 × 0.1573	7 × 0.1049	1.259	0.315	29800	1254	1049	205	83.7	16.3
FINCH	1113.0	54 × 0.1436	19 × 0.0862	1.293	0.431	39100	1430	1054	376	73.7	26.3
BUNTING	1192.5	45 × 0.1628	7 × 0.1085	1.302	0.326	32000	1342	1123	219	83.7	16.3
GRACKLE	1192.5	54 × 0.1486	19 × 0.0892	1.338	0.446	41900	1531	1128	403	73.7	26.3
SKYLARK	1272.0	36 × 0.1880	1 × 0.1880	1.316	0.1880	26400	1286	1192	94	92.7	7.3
BITTERN	1272.0	45 × 0.1681	7 × 0.1121	1.345	0.336	34100	1432	1198	234	83.7	16.3
PHEASANT	1272.0	54 × 0.1535	19 × 0.0921	1.382	0.461	43600	1634	1205	429	73.7	26.3
DIPPER	1351.5	45 × 0.1733	7 × 0.1155	1.386	0.347	36200	1521	1273	218	33.7	46.3

(Continued)

TABLE 2.1 Aluminum Conductor Steel Reinforced—Physical Properties (Continued)

Code word	Size AWG or kcmil	Stranding number & diameter (in) Aluminum	Steel	Nominal diameter (in) Complete conductor	Steel core	Rated strength (lb)	Weight (lb/1000 ft) Total	Aluminum	Steel	Percent of total weight Aluminum	Steel
MARTIN	1351.5	54 × 0.1582	19 × 0.0949	1.424	0.475	46300	1735	1279	456	73.7	26.3
BOBOLINK	1431.0	45 × 0.1783	7 × 0.1189	1.427	0.357	38300	1611	1348	263	83.7	16.3
PLOVER	1431.0	54 × 0.1628	19 × 0.0977	1.465	0.489	49100	1838	1355	483	73.7	26.3
LAPWING	1590.0	45 × 0.1880	7 × 0.1253	1.504	0.376	42200	1790	1498	292	83.8	16.3
FALCON	1590.0	54 × 0.1716	19 × 0.1030	1.545	0.515	54500	2042	1505	537	73.7	26.3
CHUKAR	1780.0	84 × 0.1456	19 × 0.0874	1.602	0.437	51000	2072	1685	387	81.3	18.7
MOCKINGBIRD	2034.5	72 × 0.1681	7 × 0.1122	1.681	0.337	46800	2163	1929	234	89.2	10.8
BLUEBIRD	2156.0	84 × 0.1602	19 × 0.0961	1.762	0.481	60300	2508	2040	468	81.3	18.7
KIWI	2167.0	72 × 0.1735	7 × 0.1157	1.735	0.347	49800	2301	2052	249	89.2	10.8
THRASHER	2312.0	76 × 0.1744	19 × 0.0814	1.802	0.407	56700	2523	2188	335	86.7	13.3
JOREA	2515.0	76 × 0.1819	19 × 0.0850	1.880	0.425	61700	2749	2383	366	86.7	13.3
High strength ACSR											
GROUSE	80.0	8 × 0.1000	1 × 0.1670	0.367	0.1670	5200	148.8	74.9	73.9	50.3	49.7
PETREL	101.8	12 × 0.0921	7 × 0.0921	0.461	0.276	10400	253.8	95.9	157.9	37.8	62.2
MINORCA	110.8	12 × 0.0961	7 × 0.0961	0.481	0.288	11300	276.3	104.4	171.9	37.8	62.2
LEGHORN	134.6	12 × 0.1059	7 × 0.1059	0.530	0.318	13600	335.5	126.8	208.7	37.8	62.2
GUINEA	159.0	12 × 0.1151	7 × 0.1151	0.576	0.345	16000	396.3	149.7	246.6	37.8	62.2
DOTTEREL	176.9	12 × 0.1214	7 × 0.1214	0.607	0.364	17300	440.9	166.6	274.4	37.8	62.2
DORKING	190.8	12 × 0.1261	7 × 0.1261	0.631	0.378	18700	475.7	179.7	296.0	37.8	62.2
COCKHIN	211.3	12 × 0.1327	7 × 0.1327	0.664	0.398	20700	526.8	199.0	327.8	37.8	62.2
BRAHMA	203.2	16 × 0.1127	19 × 0.0977	0.714	0.489	28400	674.6	191.4	483.2	28.4	71.6

Courtesy of ALCAN CABLE.
ASTM: B230, Specification for Aluminum 1350-H19 Wire for Electrical Purposes.
B232, Specification for Concentric Lay Stranded Aluminum Conductors, Coated Steel Reinforced (ACSR).
B498, Specification for Zinc-Coated (Galvanized) Steel Core Wire for Aluminum Conductors Steel Reinforced (ACSR).

TABLE 2.2 Aluminum Conductor Steel Reinforced—Electrical Properties

Code word	Size & stranding AWG or kcmil	Aluminum/ steel	Resistance DC (ohms/1000 ft) @20°C	AC 60 Hz (ohms/1000 ft) @25°C	@50°C	@75°C	60-Hz reactance 1-ft equivalent spacing Capacitive (megohms-1000 ft)	Inductive (ohms/1000 ft) @25°C	@50°C	@75°C
TURKEY	6	6/1	0.6419	0.6553	0.750	0.8159	0.7513	0.1201	0.1390	0.1439
SWAN	4	6/1	0.4032	0.4119	0.4794	0.5218	0.7149	0.1152	0.1314	0.1369
SWANATE	4	7/1	0.3989	0.4072	0.4633	0.5165	0.7102	0.11533	0.1239	0.1303
SPARROW	2	6/1	0.2534	0.2591	0.3080	0.3360	0.6785	0.1100	0.1235	0.1277
SPARATE	2	7/1	0.2506	0.2563	0.2966	0.3297	0.6737	0.1081	0.1176	0.1206
ROBIN	1	6/1	0.2011	0.2059	0.2474	0.2703	0.6600	0.1068	0.1191	0.1224
RAVEN	1/0	6/1	0.1593	0.1633	0.1972	0.2161	0.6421	0.1040	0.1138	0.1163
QUAIL	2/0	6/1	0.1265	0.1301	0.1616	0.1760	0.6241	0.1017	0.1117	0.1135
PIGEON	3/0	6/1	0.1003	0.1034	0.1208	0.1445	0.6056	0.0992	0.1083	0.1095
PENGUIN	4/0	6/1	0.0795	0.0822	0.1066	0.1157	0.5966	0.0964	0.1047	0.1053

Code word	Size & stranding AWG or kcmil	Aluminum/ steel	Resistance DC (ohms/1000 ft) @20°C	AC 60 Hz (ohms/1000 ft) @25°C	@50°C	@75°C	Capacitive (megohms-1000 ft)	Inductive (ohms/1000 ft)	GMR (ft)
WAXWING	266.8	18/1	0.0644	0.0657	0.0723	0.0788	0.576	0.0934	0.0197
PARTRIDGE	266.8	26/7	0.0637	0.0652	0.0714	0.0778	0.565	0.0881	0.0217
MERLIN	336.4	18/1	0.0510	0.0523	0.0574	0.0625	0.560	0.0877	0.0221
LINNET	336.4	26/7	0.0506	0.0517	0.0568	0.0619	0.549	0.0854	0.0244
ORIOLE	336.4	30/7	0.0502	0.0513	0.0563	0.0614	0.544	0.0843	0.0255
CHICKADEE	397.5	18/1	0.0432	0.0443	0.0487	0.0528	0.544	0.0856	0.0240
IBIS	397.5	26/7	0.0428	0.0438	0.0481	0.0525	0.539	0.0835	0.0265
LARK	397.5	30/7	0.0425	0.0434	0.0477	0.0519	0.533	0.0824	0.0277
PELICAN	477.0	18/1	0.0360	0.0369	0.0405	0.0441	0.528	0.0835	0.0263
FLICKER	477.0	24/7	0.0358	0.0367	0.0403	0.0439	0.524	0.0818	0.0283
HAWK	477.0	26/7	0.0357	0.0366	0.0402	0.0438	0.522	0.0814	0.0290
HEN	477.0	30/7	0.0354	0.0362	0.0402	0.0434	0.517	0.0803	0.0304
OSPREY	556.5	18/1	0.0309	0.0318	0.0348	0.0379	0.518	0.0818	0.0284
PARAKEET	556.5	24/7	0.0307	0.0314	0.0347	0.0377	0.512	0.0801	0.0306

(Continued)

TABLE 2.2 Aluminum Conductor Steel Reinforced—Electrical Properties (*Continued*)

Code word	AWG or kcmil	Aluminum/steel	DC (ohms/1000 ft) @20°C	AC 60 Hz @25°C	AC 60 Hz @50°C	AC 60 Hz @75°C	Capacitive (megohms-1000 ft)	Inductive (ohms/1000 ft) @25°C	GMR (ft)
DOVE	556.5	26/7	0.0305	0.0314	0.0345	0.0375	0.510	0.0795	0.0313
EAGLE	556.5	30/7	0.0300	0.0311	0.0341	0.0371	0.505	0.0786	0.0328
PEACOCK	605.0	24/7	0.0282	0.0290	0.0378	0.0347	0.505	0.0792	0.0319
SWIFT	636.0	36/1	0.0267	0.0281	0.0307	0.0334	0.509	0.0806	0.0300
KINGBIRD	636.0	18/1	0.0269	0.0278	0.0306	0.0332	0.507	0.0805	0.0301
ROOK	636.0	24/7	0.0268	0.0277	0.0300	0.0330	0.502	0.0786	0.0327
GROSBEAK	636.0	26/7	0.0267	0.0275	0.0301	0.0328	0.500	0.0780	0.0335
EGRET	636.0	30/19	0.0266	0.0273	0.0299	0.0326	0.495	0.0769	0.0351
FLAMINGO	666.6	24/7	0.0256	0.0263	0.0290	0.0314	0.498	0.0780	0.0335
STARLING	715.5	26/7	0.0238	0.0244	0.0269	0.0292	0.490	0.0767	0.0355
REDWING	715.5	30/19	0.0236	0.0242	0.0267	0.0290	0.486	0.0756	0.0372
COOT	795.0	36/1	0.0217	0.0225	0.0247	0.0268	0.492	0.0780	0.0335
TERN	795.0	45/7	0.0215	0.0225	0.0246	0.0267	0.488	0.0764	0.0352
CUCKOO	795.0	24/7	0.0215	0.0223	0.0243	0.0266	0.484	0.0763	0.0361
CONDOR	795.0	54/7	0.0215	0.0222	0.0244	0.0265	0.484	0.0759	0.0368
DRAKE	795.0	26/7	0.0214	0.0222	0.0242	0.0263	0.482	0.0756	0.0375
MALLARD	795.0	30/19	0.0213	0.0220	0.0241	0.0261	0.477	0.0744	0.0392
RUDDY	900.0	45/7	0.0191	0.0200	0.0218	0.0237	0.479	0.0755	0.0374
CANARY	900.0	54/7	0.0190	0.0197	0.0216	0.0235	0.474	0.0744	0.0392
CORNCRAKE	954.0	20/7	0.0180	0.0188	0.0206	0.0224	0.474	0.0751	0.0378
REDBIRD	954.0	24/7	0.0179	0.0186	0.0204	0.0221	0.470	0.0742	0.0396
TOWHEE	954.0	48/7	0.0180	0.0188	0.0205	0.0223	0.473	0.0745	0.0391
RAIL	954.0	45/7	0.0180	0.0188	0.0206	0.0223	0.474	0.0748	0.0385
CARDINAL	954.0	54/7	0.0179	0.0188	0.0205	0.0222	0.470	0.0757	0.0404
ORTOLAN	1033.5	45/7	0.0167	0.0175	0.0191	0.0208	0.468	0.0739	0.0401
CURLEW	1033.5	54/7	0.0165	0.0172	0.0189	0.0201	0.464	0.0729	0.0420
BLUEJAY	1113.0	45/7	0.0155	0.0163	0.0178	0.0193	0.462	0.0731	0.0416

Code Word	kcmil	Stranding							
FINCH	1113.0	54/19	0.0154	0.0161	0.0176	0.0191	0.458	0.0702	0.0436
BUNTING	1192.5	45/7	0.0144	0.0152	0.0167	0.0181	0.456	0.0723	0.0431
GRACKLE	1192.5	54/19	0.0144	0.0151	0.0165	0.0179	0.452	0.0710	0.0451
SKYLARK	1272.0	36/1	0.0135	0.0145	0.0159	0.0173	0.455	0.072	0.0427
BITTERN	1272.0	45/7	0.0135	0.0144	0.0157	0.0170	0.451	0.072	0.0445
PHEASANT	1272.0	54/19	0.0135	0.0142	0.0155	0.0169	0.447	0.070	0.0466
DIPPER	1351.5	45/7	0.0127	0.0136	0.0148	0.0161	0.447	0.071	0.0459
MARTIN	1351.5	54/19	0.0127	0.0134	0.0147	0.0159	0.442	0.070	0.0480
BOBOLINK	1431.0	45/7	0.0120	0.0129	0.0141	0.0152	0.442	0.070	0.0472
PLOVER	1431.0	54/19	0.0120	0.0127	0.0134	0.0151	0.438	0.069	0.0495
LAPWING	1590.0	45/7	0.0108	0.0117	0.0127	0.0138	0.434	0.069	0.0498
FALCON	1590.0	54/19	0.0108	0.0116	0.0126	0.0137	0.430	0.068	0.0521
CHUKAR	1780.0	84/19	0.0097	0.0106	0.0115	0.0125	0.424	0.067	0.0534
MOCKINGBIRD	2034.5	72/7	0.0085	0.0096	0.0104	0.0112	0.416	0.066	0.0553
BLUEBIRD	2156.0	84/19	0.0080	0.0090	0.0098	0.0105	0.409	0.065	0.0588
KIWI	2167.0	72/7	0.0080	0.0092	0.0099	0.0106	0.411	0.068	0.0570
THRASHER	2312.0	76/19	0.0075	0.0086	0.0092	0.0100	0.405	0.065	0.0600
JOREA	2515.0	76/19	0.0069	0.0081	0.0087	0.0093	0.399	0.064	0.0621

High strength ACSR

Code Word	kcmil	Stranding					60-Hz reactance 1-ft equivalent spacing			
							Capacitive (megohms-1000 ft)	Inductive (ohms/1000 ft)		
								@25°C	@50°C	@75°C
GROUSE	80.0	B/1	0.2065	0.2110	0.2362	0.2612	0.6547	0.1047	0.1129	0.1150
PETREL	101.8	12/7	0.1583	0.1625	0.2072	0.2394	0.6193	0.1019	0.1161	0.1282
MINORCA	110.8	12/7	0.1454	0.1491	0.1932	0.2233	0.6125	0.1017.	0.1178	0.1269
LEGHORN	134.6	12/7	0.1198	0.1233	0.1638	0.1894	0.5972	0.0998	0.1148	0.1227
GUINEA	159.0	12/7	0.1014	0.1045	0.1426	0.1653	0.5845	0.0979	0.1117	0.1189
DOTTEREL	176.9	12/7	0.0911	0.0945	0.1301	0.1513	0.5760	0.0970	0.1102	0.1169
DORKING	190.8	12/7	0.0845	0.0875	0.1229	0.1424	0.5697	0.0956	0.1093	0.1150
COCHIN	211.3	12/7	0.0763	0.0792	0.1125	0.1311	0.5618	0.0945	0.1074	0.1129
BRAHMA	203.2	16/19	0.0764	0.0790	0.1089	0.1348	0.5507	0.0934	0.1047	0.1121

Courtesy of ALCAN CABLE.

NOTES:
1. DC resistance is based on 16.946 ohn-cmil/ft 61.2% IACS for 1350 wires and 129.64 ohn-cmil/ft, 8% IACS for the steel core at 20°C with stranding increment as per ASTM B232.

TABLE 2.3 All Aluminum Conductor—Physical Properties

	Conductor size		Stranding				
Code word	AWG or kcmil	Cross-sectional area (in²)	Class	Number & dia. of strands (in)	Nominal conductor diameter (in)	Rated strength (lb)	Nominal weight (lb/1000 ft)
PEACHBELL	6	0.0206	A	7 × 0.0612	0.184	563	24.6
ROSE	4	0.0328	A	7 × 0.0772	0.232	881	39.1
IRIS	2	0.0522	AA, A	7 × 0.0974	0.292	1350	62.2
PANSY	1	0.0657	AA, A	7 × 01093	0.328	1640	78.4
POPPY	1/0	0.0829	AA, A	7 × 0.1228	0.368	1990	98.9
ASTER	2/0	0.1045	AA, A	7 × 0.1379	0.414	2510	124.8
PHLOX	3/0	0.1317	AA, A	7 × 0.1548	0.464	3040	157.2
OXLIP	4/0	0.1662	AA, A	7 × 0.1739	0.522	3830	198.4
SNEEZEWORT	250.0	0.1964	AA	7 × 0.1890	0.567	4520	234.4
VALERIAN	250.0	0.1963	A	19 × 0.1147	0.574	4660	234.3
DAISY	266.8	0.2095	AA	7 × 0.1952	0.586	4830	250.2
LAUREL	266.8	0.2095	A	19 × 0.1185	0.593	4970	250.1
PEONY	300.0	0.2358	A	19 × 0.1257	0.629	5480	281.4
TULIP	336.4	0.2644	A	19 × 0.1331	0.666	6150	315.5
DAFFODIL	350.0	0.2748	A	19 × 0.1357	0.679	6390	327.9
CANNA	397.5	0.3124	AA, A	19 × 0.1447	0.724	7110	372.9
GOLDENTUFT	450.0	0.3534	AA	19 × 0.1539	0.769	7890	421.8
COSMOS	477.0	0.3744	AA	19 × 0.1584	0.792	8360	448.8
SYRINGA	477.0	0.3744	A	37 × 0.1135	0.795	8690	446.8
ZINNIA	500.0	0.3926	AA	19 × 0.1622	0.811	8760	468.5
DAHLIA	556.5	0.4368	AA	19 × 0.1711	0.856	9750	521.4
MISTLETOE	556.5	0.4368	A	37 × 0.1226	0.858	9940	521.3
MEADOWSWEET	600.0	0.4709	AA, A	37 × 0.1273	0.891	10700	562.0
ORCHID	636.0	0.4995	AA, A	37 × 0.1311	0.918	11400	596.0
HEUCHERA	650.0	0.5102	AA	37 × 0.1325	0.928	11600	609.8
VERBENA	700.0	0.5494	AA	37 × 0.1375	0.963	12500	655.7
VIOLET	715.5	0.5623	AA	37 × 0.1391	0.974	12800	671.0
NASTURTIUM	715.5	0.5619	A	61 × 0.1083	0.975	13100	671.0

PETUNIA	750.0	0.5893	AA	37 × 0.1424	0.997	13100	703.2
ARBUTUS	795.0	0.6245	AA	37 × 0.1466	1.026	13900	745.3
LILAC	795.0	0.6248	A	61 × 0.1142	1.028	14300	745.7
COCKSCOMB	900.0	0.7072	AA	37 × 0.1560	1.092	16400	844.0
MAGNOLIA	954.0	0.7495	AA	37 × 0.1606	1.124	16400	894.5
GOLDENROD	954.0	0.7498	A	61 × 0.1251	1.126	16900	894.8
HAWKWEED	1000.0	0.7854	AA	37 × 0.1644	1.151	17200	937.3
BLUEBELL	1033.5	0.8114	AA	37 × 0.1671	1.170	17700	968.4
LARKSPUR	1033.5	0.8122	A	61 × 0.1302	1.172	18300	969.2
MARIGOLD	1113.0	0.8744	AA, A	61 × 0.1351	1.216	19700	1044
HAWTHORN	1192.5	0.9363	AA, A	61 × 0.1398	1.258	21100	1117
NARCISSUS	1272.0	0.9990	AA, A	61 × 0.1444	1.300	22000	1192
COLUMBINE	1351.5	1.061	AA, A	61 × 0.1488	1.340	23400	1266
CARNATION	1431.0	1.124	AA, A	61 × 0.1532	1.379	24300	1342
COREOPSIS	1590.0	1.248	AA	61 × 0.1614	1.454	27000	1489
JESSAMINE	1750.0	1.375	AA	61 × 0.1694	1.525	29700	1641
COWSLIP	2000.0	1.570	A	91 × 0.1482	1.630	34200	1873
LUPINE	2500.0	1.962	A	91 × 0.1657	1.823	41900	2365
TRILLIUM	3000.0	2.356	A	127 × 0.1537	1.998	50300	2840
BLUEBONNET	3500.0	2.749	A	127 × 0.1660	2.158	58700	3345

Courtesy of ALCAN CABLE.
ASTM B230: Specification for Aluminum 1350-H19 Wire for Electrical Purposes.
B231: Specification for Concentric-Lay-Stranded Aluminum 1350 Conductors.

TABLE 2.4 All Aluminum Conductor—Electrical Properties

Code word	AWG or kcmil	Number of strands	DC 20°C (ohms/ 1000 ft)	AC 60 Hz — 25°C (ohms/ 1000 ft)	AC 60 Hz — 50°C (ohms/ 1000 ft)	AC 60 Hz — 75°C (ohms/ 1000 ft)	Capacitive (megohms 1000 ft)	Inductive (ohms/ 1000 ft)	GMR (ft)
PEACHBELL	6	7	0.6593	0.6725	0.7392	0.8059	0.7660	0.1193	0.00555
ROSE	4	7	0.4144	0.4227	0.4645	0.5064	0.7296	0.1140	0.00700
IRIS	2	7	0.2602	0.2655	0.2929	0.3182	0.6929	0.1087	0.00883
PANSY	1	7	0.2066	0.2110	0.2318	0.2527	0.6716	0.1061	0.00991
POPPY	1/0	7	0.1638	0.1671	0.1837	0.2002	0.6550	0.1034	0.0111
ASTER	2/0	7	0.1299	0.1326	0.1456	0.1587	0.6346	0.1008	0.0125
PHLOX	3/0	7	0.1031	0.1053	0.1157	0.1259	0.6188	0.0981	0.0140
OXLIP	4/0	7	0.0817	0.0835	0.0917	0.1000	0.6029	0.0955	0.0158
SNEEZEWORT	250.0	7	0.0691	0.0706	0.0777	0.0847	0.586	0.0934	0.0171
VALERIAN	250.0	19	0.0691	0.0706	0.0777	0.0847	0.586	0.0922	0.0181
DAISY	266.8	7	0.0648	0.0663	0.0727	0.0794	0.581	0.0926	0.0177
LAUREL.	266.8	19	0.0648	0.0663	0.0727	0.0794	0.581	0.0915	0.0187
PEONY	300.0	19	0.0575	0.0589	0.0648	0.0705	0.570	0.0902	0.0198
TULIP	336.4	19	0.0513	0.0527	0.0578	0.0629	0.560	0.0888	0.0210
DAFFODIL	350.0	19	0.0494	0.0506	0.0557	0.0606	0.560	0.0883	0.0214
CANNA	397.5	19	0.0435	0.0445	0.0489	0.0534	0.549	0.0869	0.0228
GOLDENTUFT	450.0	19	0.0384	0.0394	0.0434	0.0472	0.539	0.0854	0.0243
COSMOS	477.0	19	0.0363	0.0373	0.0409	0.0445	0.533	0.0848	0.0250
SYRINGA	477.0	37	0.0363	0.0373	0.0409	0.0445	0.533	0.0845	0.0254
ZINNIA	500.0	19	0.0346	0.0356	0.0390	0.0426	0.531	0.0843	0.0256
DAHLIA	556.5	19	0.0311	0.0320	0.0352	0.0383	0.522	0.0830	0.0270
MISTLETOE	556.5	37	0.0311	0.0320	0.0352	0.0383	0.522	0.0826	0.0275
MEADOWSWEET	600.0	37	0.0288	0.0297	0.0326	0.0356	0.516	0.0818	0.0285
ORCHID	636.0	37	0.0272	0.0282	0.0309	0.0335	0.511	0.0811	0.0294
HEUCHERA	650.0	37	0.0266	0.0275	0.0301	0.0324	0.510	0.0808	0.0297
VERBENA	700.0	37	0.0247	0.0256	0.0280	0.0305	0.504	0.0799	0.0308

VIOLET	715.5	37	0.0242	0.0252	0.0275	0.0299	0.502	0.0797	0.0312
NASTURTIUM	715.5	61	0.0242	0.0252	0.0275	0.0299	0.502	0.0795	0.0314
PETUNIA	750.0	37	0.0230	0.0251	0.0263	0.0286	0.498	0.0792	0.0319
ARBUTUS	795.0	37	0.0217	0.0227	0.0248	0.0269	0.494	0.0780	0.0328
LILAC	795.0	61	0.0217	0.0227	0.0248	0.0269	0.494	0.0784	0.0331
COCKSCOMB	900.0	37	0.0192	0.0201	0.0220	0.0239	0.484	0.0771	0.0349
MAGNOLIA	954.0	37	0.0181	0.0191	0.0208	0.0227	0.479	0.0763	0.0360
GOLDENROD	954.0	61	0.0181	0.0191	0.0208	0.0227	0.479	0.0763	0.0362
HAWKWEED	1000.0	37	0.0173	0.0192	0.0199	0.0216	0.476	0.0759	0.0368
BLUEBELL	1033.5	37	0.0167	0.0177	0.0193	0.0210	0.473	0.0756	0.0374
LARKSPUR	1033.5	61	0.0167	0.0177	0.0193	0.0210	0.473	0.0754	0.0377
MARIGOLD	1113.0	61	0.0155	0.0165	0.0180	0.0195	0.467	0.0744	0.0391
HAWTHORN	1192.5	61	0.0145	0.0155	0.0169	0.0183	0.462	0.0737	0.0405
NARCISSUS	1272.0	61	0.0136	0.0146	0.0159	0.0173	0.457	0.0729	0.0418
COLUMBINE	1351.5	61	0.0128	0.0138	0.0151	0.0163	0.452	0.0722	0.0431
CARNATION	1431.0	61	0.0121	0.0132	0.0143	0.0155	0.447	0.0715	0.0444
COREOPSIS	1590.0	61	0.0109	0.0120	0.0130	0.0141	0.439	0.0705	0.0468
JESSAMINE	1750.0	61	0.0099	0.0111	0.0120	0.0129	0.432	0.0693	0.0490
COWSLIP	2000.0	91	0.0087	0.0099	0.0107	0.0115	0.421	0.0677	0.0525
LUPINE	2500.0	91	0.0070	0.0084	0.0091	0.0097	0.404	0.0652	0.0588
TRILLIUM	3000.0	127	0.0058	0.0074	0.0079	0.0084	0.389	0.0629	0.0646
BLUEBONNET	3500.0	127	0.0050	0.0068	0.0072	0.0076	0.378	0.0612	0.0697

Courtesy of ALCAN CABLE.

NOTES:
1. DC resistance is based on 16.946 ohm-cmil/ft, (c)20° C, 61.2% IACS with stranding increments as per ASTM B231.

TABLE 2.5 Aluminum Conductor Steel Supported—Physical Properties

Code word	kcmil	Stranding number & diameter		Complete diameter (in)	Steel care (in)	ACSS/MA Rated strength (lb)	Weight (lb/100 ft)		
		Aluminum	Steel				Total	Aluminum	Steel
PARTRIDGE/ACSS	266.8	26 × 0.1013	7 × 0.0788	0.642	0.236	8,880	366.9	251.3	115.8
LINNET/ACSS	336.4	26 × 0.1137	7 × 0.0884	0.720	0.265	11,200	462.0	316.6	145.4
ORIOLE/ACSS	336.4	30 × 01059	7 × 0.1059	0.741	0.317	14,800	526.4	317.7	208.7
IBIS/ACSS	397.5	26 × 0.1236	7 × 0.0961	0.783	0.288	13,000	546.0	374.1	171.9
LARK/ACSS	397.5	30 × 0.1151	7 × 0.1151	0.806	0.345	17,500	621.8	375.3	246.5
FLICKER/ACSS	477	24 × 0.1410	7 × 0.0940	0.846	0.282	13,000	613.9	449.4	164.5
HAWK/ACSS	477	26 × 0.1354	7 × 0.1053	0.858	0.316	15,600	655.3	449.0	206.3
HEN/ACSS	477	30 × 0.1261	7 × 0.1261	0.883	0.378	21,000	746.4	450.4	296.0
PARAKEET/ACSS	556.5	24 × 0 1523	7 × 0.1015	0.914	0.304	15,200	716.1	524.3	191.8
DOVE/ACSS	556.5	26 × 0.1463	7 × 0.1138	0.927	0.341	18,200	765.2	524.2	241.0
EAGLE/ACSS	556.5	30 × 0.1362	7 × 0.1362	0.953	0.409	24,500	870.7	525.5	345.2
PEACOCK/ACSS	605	24 × 0.1588	7 × 0.1059	0.953	0.318	16,500	778.8	570.1	208.7
ROOK/ACSS	636	24 × 0.1628	7 × 0.1085	0.977	0.326	17,300	818.2	599.1	219.1
GROSBEAK/ACSS	636	26 × 0.1564	7 × 0.1216	0.990	0.365	20,700	874.2	599.0	275.2
EGRET/ACSS	636	30 × 0.1456	19 × 0.0874	1.019	0.437	28,000	987.2	600.5	386.7
FLAMINGO/ACSS	666.6	24 × 0.1667	7 × 0.1111	1.000	0.333	18,200	857.9	628.2	229.7
STARLING/ACSS	715.5	26 × 0.1659	7 × 0.1290	1.051	0.387	23,300	983.7	674.0	309.7
REDWING/ACSS	715.5	30 × 0.1544	19 × 0.0926	1.081	0.463	30,800	1109.3	675.3	434.0
TERN/ACSS	795	45 × 0.1329	7 × 0.0886	1.063	0.266	14,200	895	748.6	146.4
CUCKOO/ACSS	795	24 × 0.1820	7 × 0.1213	1.092	0.364	21,700	1023	748.8	274.2
CONDOR/ACSS	795	54 × 0.1213	7 × 0.1213	1.092	0.364	21,700	1022	748.4	273.6
DRAKE/ACSS	795	26 × 0.1749	7 × 0.1360	1.108	0.408	25,900	1093	749.1	343.9
MALLARD/ACSS	795	30 × 0.1628	19 × 0.0977	1.140	0.489	34,300	1233.9	750.7	483.2
RUDDY/ACSS	900	45 × 0.1414	7 × 0.0943	1.131	0.283	15,800	1013	847.4	165.6
CANARY/ACSS	900	54 × 0.1291	7 × 0.1291	1.162	0.387	24,600	1158	847.7	310.3
RAIL/ACSS	954	45 × 0.1456	7 × 0.0971	1.165	0.291	16,700	1074	898.5	175.5

CARDINAL/ACSS	954	54 × 0.1329	7 × 0.1329	1.196	0.399	26,000	1227.1	898.3	328.8
ORTOLAN/ACSS	1033.5	45 × 0.1515	7 × 0.1010	1.212	0.303	18,100	1163	972.8	190.2
CURLEW/ACSS	1033.5	54 × 0.1383	7 × 0.1383	1.246	0.415	28,200	1329	972.8	356.2
BLUEJAY/ACSS	1113	45 × 0.1573	7 × 0.1049	1.259	0.315	19,500	1254	1048.7	205.3
FINCH/ACSS	1113	54 × 0.1436	19 × 0.0862	1.293	0.431	30,400	1430	1053.9	376.1
BUNTING/ACSS	1192.5	45 × 0.1628	7 × 0.1085	1.302	0.326	21,400	1142	1123.0	219.0
GRACKLE/ACSS	1192.5	54 × 0.1486	19 × 0.0892	1.338	0.446	32,600	1531	1128.6	402.4
BITTERN/ACSS	1272	45 × 0.1681	7 × 0.1121	1.345	0.336	22,300	1432	1197.7	234.3
PHEASANT/ACSS	1272	54 × 0.1535	19 × 0.0921	1.382	0.461	34,100	1634	1204.3	429.7
DIPPER/ACSS	1351.5	45 × 0.1733	7 × 0.1155	1.386	0.347	23,700	1521	1272.9	248.1
MARTIN/ACSS	1351.5	54 × 0.1582	19 × 0.0949	1.424	0.475	36,200	1735	1279.2	455.8
BOBOLINK/ACSS	1431	45 × 0.1783	7 × 0.1189	1.427	0.357	25,100	1611	1347.5	263.5
PLOVER/ACSS	1431	54 × 0.1628	19 × 0.0977	1.465	0.489	38,400	1838	1354.6	483.4
LAPWING/ACSS	1590	45 × 0.1880	7 × 0.1253	1.504	0.376	27,900	1790	1498.1	291.9
FALCON/ACSS	1590	54 × 0.1716	19 × 0.1030	1.545	0.515	42,600	2042	1505.0	537.0
CHUKAR/ACSS	1780	84 × 0.1456	19 × 0.0874	1.602	0.437	35,400	2072	1685.5	386.5
BLUEBIRD/ACSS	2156	84 × 0.1602	19 × 0.0961	1.762	0.481	42,100	2508	2040.4	467.6
KIWI/ACSS	2167	72 × 0.1735	7 × 0.1157	1.735	0.347	29,000	2301	2051.4	249.6
THRASHER/ACSS	2312	71 × 0.1744	19 × 0.0814	1.802	0.407	35,600	2523	2187.9	335.1

Courtesy of ALCAN CABLE.
ASTM: Conductors manufactured to ASTM B856, Specification for Concentric-Lay-Stranded Aluminum Conductors, Coated Steel Supported 3ACSS).
ACSS: Aluminum Conductor, Steel Supported.
ACSS/MA, supported with ZN-5A1-MM coated steel core wire, coating Class A in accordance with specification ASTM B802.

TABLE 2.6 Aluminum Conductor Steel Supported—Electrical Properties

Code word	Size & stranding kcmil	Stranding	Resistance (others per 1000 ft) DC @20°C	AC 60 Hz @25°C	@75°C	Reactance Capacitive (megohms-1000 ft)	Inductive (ohms/1000 ft)	GMR (ft)	Ampacity
PARTRIDGE/ACSS	266.8	26/7	0.0619	0.0633	0.0760	0.565	0.088	0.0217	814
LINNET/ACSS	336.4	26/7	0.0492	0.0502	0.0605	0.549	0.085	0.0244	943
ORIOLE/ACSS	336.4	30/7	0.0488	0.0498	0.0600	0.544	0.084	0.0255	955
IBIS/ACSS	397.5	26/7	0.0416	0.0426	0.0513	0.539	0.084	0.0265	1051
LARK/ACSS	397.5	30/7	0.0413	0.0422	0.0507	0.533	0.082	0.0277	1069
FLICKER/ACSS	477	24/7	0.0348	0.0357	0.0429	0.524	0.082	0.0283	1180
HAWK/ACSS	477	26/7	0.0347	0.0356	0.0428	0.522	0.081	0.0290	1186
HEN/ACSS	477	30/7	0.0344	0.0352	0.0424	0.517	0.080	0.3040	1200
PARAKEET/ACSS	556.5	24/7	0.0298	0.0305	0.0368	0.512	0.080	0.0306	1300
DOVE/ACSS	556.5	26/7	0.0296	0.0305	0.0366	0.510	0.080	0.0313	1316
EAGLE/ACSS	556.5	30/7	0.0291	0.0302	0.0362	0.505	0.079	0.0328	1336
PEACOCK/ACSS	605	24/7	0.0274	0.0282	0.0339	0.505	0.079	0.0319	1378
ROOK/ACSS	636	24/7	0.0260	0.0269	0.0322	0.502	0.079	0.0327	1429
GROSBEAK/ACSS	636	26/7	0.0259	0.0267	0.0320	0.500	0.078	0.0335	1438
EGRET/ACSS	636	30/19	0.0258	0.0265	0.0319	0.495	0.077	0.0351	1449
FLAMINGO/ACSS	666.6	24/7	0.0249	0.0256	0.0307	0.498	0.078	0.0335	1473
STARLING/ACSS	715.5	26/7	0.0231	0.0237	0.0285	0.490	0.077	0.0355	1550
REDWING/ACSS	715.5	30/19	0.0229	0.0235	0.0283	0.486	0.076	0.0372	1568
TERN/ACSS	795	45/7	0.0210	0.0219	0.0261	0.488	0.076	0.0952	1636
CUCKOO/ACSS	795	24/7	0.0209	0.0217	0.0260	0.484	0.076	0.0361	1647
CONDOR/ACSS	795	54/7	0.0209	0.0216	0.0259	0.484	0.076	0.0368	1650
DRAKE/ACSS	795	26/7	0.0210	0.0216	0.0259	0.482	0.076	0.0375	1660
MALLARD/ACSS	795	30/19	0.0207	0.0214	0.0255	0.477	0.074	0.0392	1693
RUDDY/ACSS	900	45/7	0.0186	0.0194	0.0232	0.479	0.076	0.0374	1766
CANARY/ACSS	900	54/7	0.0185	0.0191	0.0230	0.474	0.074	0.0392	1780
RAIL/ACSS	954	45/7	0.0172	0.0180	0.0215	0.474	0.075	0.0385	1844

CARDINAL/ACSS	954	54/7	0.0174	0.0181	0.0217	0.470	0.076	0.0404	1856
ORTOLAN/ACSS	1033.5	45/7	0.0162	0.0170	0.0203	0.468	0.074	0.0401	1933
CURLEW/ACSS	1033.5	54/7	0.0160	0.0167	0.0202	0.464	0.073	0.0420	1936
BLUEJAY/ACSS	1113	45/7	0.0151	0.0158	0.0189	0.462	0.073	0.0416	2025
FINCH/ACSS	1113	54/19	0.0150	0.0156	0.0190	0.458	0.070	0.0436	2020
BUNTING/ACSS	1192.5	45/7	0.0140	0.0148	0.0177	0.456	0.072	0.0431	2116
GRACKLE/ACSS	1192.5	54/19	0.0140	0.0147	0.0177	0.452	0.071	0.0451	2122
BITTERN/ACSS	1272	45/7	0.0131	0.0140	0.0166	0.451	0.072	0.0445	2223
PHEASANT/ACSS	1272	54/19	0.0131	0.0138	0.0166	0.447	0.070	0.0466	2221
DIPPER/ACSS	1351.5	45/7	0.0123	0.0132	0.0157	0.447	0.071	0.0459	2303
MARTIN/ACSS	1431	54/19	0.0123	0.0130	0.0156	0.442	0.070	0.0480	2313
BOBOLINK/ACSS	1431	45/7	0.0117	0.0125	0.0149	0.442	0.070	0.0472	2383
PLOVER/ACSS	1590	54/19	0.0117	0.0123	0.0148	0.438	0.069	0.0495	2395
LAPWING/ACSS	1590	45/7	0.0105	0.0114	0.0135	0.434	0.069	0.0498	2559
FALCON/ACSS	1590	54/19	0.0105	0.0113	0.0136	0.430	0.068	0.0521	2543
CHUKAR/ACSS	1780	84/19	0.0094	0.0103	0.0122	0.424	0.067	0.0534	2749
BLUEBIRD/ACSS	2156	84/19	0.0078	0.0087	0.0103	0.409	0.065	0.0588	3089
KIWI/ACSS	2167	72/7	0.0078	0.0089	0.0104	0.411	0.068	0.0570	3089
THRASHER/ACSS	2312	76/19	0.0073	0.0084	0.0098	0.405	0.065	0.0600	3227

Courtesy of ALCAN CABLE.
NOTES:
1. Ampacity based on a 200°C conductor temperature, 25°C ambient temperature, 2 ft/sec, wind, in sun, with emissivity of 0.5 and a coefficient of solar absorption of 0.5, at sea level.
2. Resistance and ampacity based on 63% IACS Al and 8% IACS steel core wire @20°C.

TABLE 2.7 All Aluminum Alloy Conductor—Physical and Electrical Properties

Code word	Conductor size (kcmil)	Conductor area (in²)	Stranding number and diameter (in)	Nominal diameter (in)	ACSR with equal diameter		Rated strength (lb)	Nominal weight (lb/1000 ft)
					Size	Stranding		
AKRON	30.58	0.024	7 × 0.0661	0.198	6	6/1	1.110	28.5
ALTON	48.69	0.0382	7 × 0.0834	0.250	4	6/1	1.760	45.4
AMES	77.47	0.0608	7 × 0.1052	0.316	2	6/1	2.800	72.2
AZUSA	123.3	0.0968	7 × 0.1327	0.398	1/0	6/1	4.270	114.9
ANAHEIM	155.4	0.1221	7 × 0.1490	0.447	2/0	6/1	5.390	144.9
AMHERST	195.7	0.1537	7 × 0.1672	0.502	3/0	6/1	6.790	182.5
ALLIANCE	246.9	0.1939	7 × 0.1878	0.563	4/0	6/1	8.560	230.2
BUTTE	312.8	0.2456	19 × 0.1283	0.642	266.8	26/7	10.500	291.6
CANTON	394.5	0.3099	19 × 0.1441	0.721	336.4	26/7	13.300	367.9
CAIRO	465.4	0.3655	19 × 0.1565	0.783	397.5	26/7	15.600	433.9
DARIEN	559.5	0.4394	19 × 0.1716	0.858	477	26/7	18.800	521.7
ELGIN	652.4	0.5124	19 × 0.1853	0.927	556.5	26/7	21.900	608.3
FLINT	740.8	0.5818	37 × 0.1415	0.991	636	26/7	24.400	690.8
GREELEY	927.2	0.7282	37 × 0.1583	1.108	795	26/7	30.500	864.6

Code word	Conductor size (kcmil)	Stranding number and diameter (in)	Approx. AAC size of equivalent resistance	DC 20°C (ohms/ 1000 ft)	AC 60 Hz Resistance 25°C (ohms/ 1000 ft)	50°C (ohms/ 1000 ft)	75°C (ohms/ 1000 ft)	GMR (ft)	Neutral 60-Hz reactance 1-ft spacing Inductive (ohm/ 1000 ft)	Capacitive (megohms- 1000 ft)
AKRON	30.58	7 × 0.0661	6	0.6589	0.670	0.727	0.784	0.00599	0.118	0.751
ALTON	48.69	7 × 0.0834	4	0.4138	0.420	0.456	0.492	0.00756	0.112	0.715
AMES	77.47	7 × 0.1052	2	0.2600	0.265	0.288	0.311	0.00954	0.107	0.678
AZUSA	123.3	7 × 0.1327	1/0	0.1635	0.166	0.180	0.195	0.0120	0.102	0.642
ANAHEIM	155.4	7 × 0.1490	2/0	0.1297	0.132	0.143	0.155	0.0135	0.0989	0.624
AMHERST	195.7	7 × 0.1672	3/0	0.1030	0.105	0.114	0.123	0.0152	0.0963	0.606
ALLIANCE	246.9	7 × 0.1878	4/0	0.0816	0.0831	0.0902	0.0973	0.0170	0.0936	0.588
BUTTE	312.8	19 × 0.1283	266.8	0.0644	0.0657	0.0712	0.0769	0.0202	0.0896	0.567
CANTON	394.5	19 × 0.1441	336.4	0.0511	0.0523	0.0566	0.0610	0.0227	0.0870	0.549
CAIRO	465.4	19 × 0.1565	397.5	0.0433	0.0443	0.0481	0.0517	0.0247	0.0851	0.536
DARIEN	559.5	19 × 0.1716	477.0	0.0360	0.0369	0.0400	0.0431	0.0271	0.0829	0.522
ELGIN	652.4	19 × 0.1853	556.5	0.0309	0.0318	0.0345	0.0371	0.0292	0.0812	0.510
FLINT	740.8	37 × 0.1415	636.0	0.0272	0.0280	0.0305	0.0328	0.0317	0.0793	0.499
GREELEY	927.2	37 × 0.1583	795.0	0.0217	0.0225	0.0244	0.0263	0.0354	0.0768	0.482

Courtesy of ALCAN CABLE.
ASTM: B398, Specification for Aluminum Alloy 6201-T81 Wire for Electrical Purposes.
B399, Specification for Concentric Lay-Stranded Aluminum Alloy 6201-T81 Conductors.
NOTES:
1. DC resistance is based on 19.755 ohm-cmil/ft @20°C (68°F) 52.5% IACS. With standard stranding increment of 2%.

TABLE 2.8 Aluminum Conductor Alloy Reinforced—Physical and Electrical Properties

Size kcmil	Stranding number & diameter (in) 1350	Stranding number & diameter (in) 6201	Cross-sectional area (in²) 1350	Cross-sectional area (in²) 6201	Cross-sectional area (in²) Total	Nominal conductor diameter (in)	ACSR with similar diameter	Nominal weight (lb/1000 ft)	Rated strength (lb)
503.6	15 × .1628	4 × .1628	0.3122	0.0833	0.3955	0.814	PELICAN	473	10,500
587.2	15 × .1758	4 × .1758	0.3641	0.0971	0.4612	0.879	OSPREY	551	12,200
649.5	18 × .1325	19 × .1325	0.2482	0.2620	0.5102	0.927	DOVE	608	15,600
653.1	12 × .1854	7 × .1854	0.3240	0.1890	0.5130	0.927	DOVE	612	15,400
739.8	18 × .1414	19 × .1414	0.2827	0.2983	0.5810	0.990	GROSBEAK	693	18,800
853.7	30 × .1519	7 × 1519	0.5437	0.1268	0.6705	1.063	TERN	801	17,500
853.7	24 × .1519	13 × .1519	0.4349	0.2356	0.6705	1.063	TERN	800	19,300
927.2	24 × .1583	13 × .1583	0.4723	0.2559	0.7282	1.108	DRAKE	869	20,900
1024.5	30 × .1664	7 × .1664	0.6524	0.1522	0.8046	1.165	RAIL	961	20,900
1024.5	24 × .1664	13 × 1664	0.5219	0.2827	0.8046	1.165	RAIL	961	23,100
1080.6	24 × .1709	13 × 1709	0.5505	0.2982	0.8487	1.196	CARDINAL	1013	24,400
1080.6	18 × .1709	19 × 1709	0.4129	0.4358	0.8487	1.196	CARDINAL	1012	27,200
1109.0	30 × .1731	7 × .1731	0.7060	0.1647	0.8707	1.212	ORTOLAN	1041	22,700
1109.0	24 × .1731	13 × .1731	0.5648	0.3059	0.8707	1.212	ORTOLAN	1040	25,000
1172.0	30 × .1780	7 × .1780	0.7465	0.1742	0.9207	1.246	CURLEW	1100	24,000
1172.0	18 × .1780	19 × .1780	0.4479	0.4728	0.9207	1.246	CURLEW	1098	29,500
1198.0	30 × .1799	7 × .1799	0.7626	0.1779	0.9405	1.259	BLUEJAY	1124	24,500
1198.0	24 × .1799	13 × .1799	0.6101	0.3304	0.9405	1.259	BLUEJAY	1123	27,100
1277.0	54 × .1447	7 × .1447	0.8880	0.1151	1.0031	1.302	BUNTING	1199	24,600
1277.0	42 × .1447	19 × .1447	0.6907	0.3124	1.0031	1.302	BUNTING	1198	28,400
1361.5	54× .1494	7 × .1494	0.9466	0.1227	1.0693	1.345	BITTERN	1278	26,300
1534.4	42× .1586	19 × .1586	0.8297	0.3754	1.2051	1.427	BOBOLINK	1439	33,800
1703.0	48 × .1671	13 × .1671	1.0527	0.2851	1.3378	1.504	LAPWING	1598	34,600
1798.0	42 × .1717	19 × .1717	0.9725	0.4399	1.4124	1.545	FALCON	1686	39,600
1933.0	42 × .1780	19 × .1780	1.0452	0.4728	1.5180	1.602	CHUKAR	1813	42,500
2338.0	42 × .1958	19 × .1958	1.2646	0.5721	1.8367	1.762	BLUE8IRD	2214	51,500
2338.0	48 × .1958	13 × .1958	1.4453	0.3914	1.8367	1.762	BLUEBIRD	2215	47,500
2493.0	54 × .1655	37 × .1655	1.1617	0.7959	1.9576	1.821	KINGFISHER	2358	57,600
2493.0	72 × .1655	19 × .1655	1.5489	0.4087	1.9576	1.821	KINGFISHER	2362	50,400

Size kcmil	Stranding		Resistance DC 20°C (ohms/ 1000 ft)	AC 60 Hz 25°C (ohms/ 1000 ft)	50°C (ohms/ 1000 ft)	75°C (ohms/ 1000 ft)	60-Hz reactance 1-ft equivalent spacing Capacitive (megohms 1000 ft)	Inductive (ohms/ 1000 ft)	GMR (ft)
	1350	6201							
503.6	15 × .1628	4 × .1628	0.0354	0.0364	0.0398	0.0433	0.531	0.0841	0.0257
587.2	15 × 1758	4 × .1758	0.0303	0.0312	0.0342	0.0371	0.518	0.0824	0.0277
649.5	18 × .1325	19 × .1325	0.0287	0.0295	0.0322	0.0349	0.509	0.0812	0.0292
653.1	12 × .1854	7 × .1854	0.0279	0.0288	0.0315	0.0341	0.509	0.0811	0.0293
739.8	18 × .1414	19 × .1414	0.0252	0.0259	0.0283	0.0307	0.499	0.0793	0.0317
853.7	30 × .1519	7 × .1519	0.0208	0.0216	0.0235	0.0257	0.488	0.0777	0.0340
853.7	24 × .1519	13 × .1519	0.0213	0.0222	0.0242	0.0262	0.488	0.0777	0.0340
927.2	24 × .1583	13 × .1583	0.0208	0.0216	0.0236	0.0252	0.482	0.0767	0.0355
1024.5	30 × .1664	7 × .1664	0.0173	0.0182	0.0199	0.0215	0.474	0.0756	0.0373
1024.5	24 × .1664	13 × .1664	0.0178	0.0186	0.0203	0.0219	0.474	0.0756	0.0373
1080.6	24 × .1709	13 × .1709	0.0168	0.0176	0.0192	0.0208	0.470	0.0750	0.0383
1080.6	18 × .1709	19 × .1709	0.0172	0.0181	0.0196	0.0213	0.470	0.0750	0.0383
1109.0	30 × .1731	7 × .1731	0.0160	0.0169	0.0184	0.0199	0.468	0.0747	0.0388
1109.0	24 × .1731	13 × .1731	0.0164	0.0172	0.0187	0.0203	0.468	0.0747	0.0388
1172.0	30 × .1780	7 × .1780	0.0152	0.0160	0.0174	0.0189	0.463	0.0740	0.0399
1172.0	18 × .1780	19 × .1780	0.0159	0.0166	0.0181	0.0195	0.463	0.0740	0.0399
1198.0	30 × .1799	7 × .1799	0.0149	0.0155	0.0170	0.0184	0.462	0.0738	0.0403

(Continued)

TABLE 2.8 Aluminum Conductor Alloy Reinforced—Physical and Electrical Properties (Continued)

Size kcmil	Stranding 1350	Stranding 6201	Resistance DC 20°C (ohms/1000 ft)	Resistance AC 60 Hz 25°C (ohms/1000 ft)	50°C (ohms/1000 ft)	75°C (ohms/1000 ft)	60 Hz reactance 1-ft equivalent spacing Capacitive (megohms 1000 ft)	Inductive (ohms/1000 ft)	GMR (ft)
1198.0	24 × .1799	13 × .1799	0.0152	0.0159	0.0173	0.0188	0.462	0.0738	0.0403
1277.0	54 × .1447	7 × .1447	0.0138	0.0149	0.0161	0.0174	0.456	0.0729	0.0419
1277.0	42 × .1447	19 × .1447	0.0142	0.0152	0.0165	0.0178	0.456	0.0729	0.0419
1361.5	54 × .1494	7 × .1494	0.0129	0.0138	0.0151	0.0163	0.451	0.0721	0.0433
1534.4	42 × .1586	19 × .1586	0.0118	0.0127	0.0139	0.0152	0.442	0.0708	0.0459
1703.0	48 × .1671	13 × .1671	0.0105	0.0115	0.0125	0.0135	0.434	0.0696	0.0484
1798.0	42 × .1717	19 × .1717	0.0101	0.0110	0.0119	0.0128	0.430	0.0690	0.0497
1933.0	42 × .1780	19 × .1780	0.0094	0.0102	0.0113	0.0122	0.424	0.0682	0.0515
2338.0	42 × .1958	19 × .1958	0.0078	0.0089	0.0096	0.0103	0.409	0.0660	0.0567
2338.0	48 × .1958	13 × .1958	0.0077	0.0088	0.0095	0.0102	0.409	0.0660	0.0567
2493.0	54 × .1655	37 × .1655	0.0074	0.0087	0.0093	0.0100	0.404	0.0652	0.0587
2493.0	72 × .1655	19 × .1655	0.0072	0.0085	0.0090	0.0098	0.404	0.0652	0.0587

Courtesy of ALCAN CABLE.
ASTM: B230, Specification for Aluminum 1350-H19 Wire for Electrical Purposes;
B398, Specification for Aluminum Alloy 6201-T81 Wire for Electrical Purposes;
B524, Specification for Concentric Lay-Stranded Aluminum Conductors, Aluminum Alloy Reinforced (ACAR, 1350/6201).
NOTES:
1. DC resistance is based on 16.946 ohm-cmil/ft for 1350 wires at 61.2% IACS and 19.755 ohm-cmil/ft for 6201 wires at 52.5%.
2. IACS @20°C, with standard increments as per ASTM B524.

Wire Sizes

American wire gauge.   A numbering system to classify wire size based upon the number of steps required to draw the wire as it is being manufactured. The greater the gage number of wire, the smaller the wire (Fig. 2.6).

Circular mil.   The area contained in a circle having a diameter of $^1/_{1000}$ inch (Fig. 2.7).

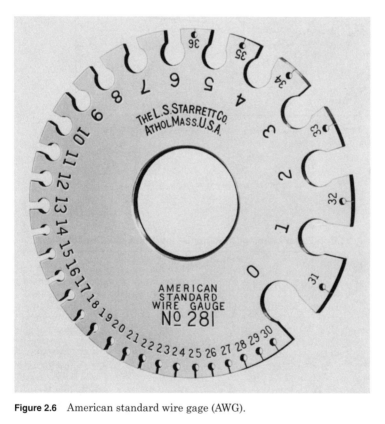

Figure 2.6   American standard wire gage (AWG).

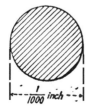

Figure 2.7   One circular mil is an area of circle whose diameter is $^1/_{1000}$ in.

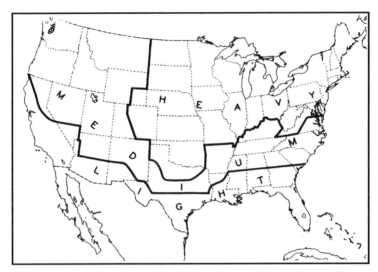

Figure 2.8 General loading map of the United States. (*Courtesy of National Electrical Safety Code.*)

TABLE 2.9   District Conductor Loading Conditions—Ice, Wind, and Temperature

	Loading districts (for use with Rule 250 B)			Extreme wind loading (for use with Rule 250C)	Extreme ice loading with concurrent wind (for use with Rule 250D)
	Heavy	Medium	Light		
Radial thickness of ice (mm)	12.5	6.5	0	0	See Figure 250-3
(in)	0.50	0.25	0	0	See Figure 250-3
Horizontal wind pressure (Pa)	190	190	430	See Figure 250-2	See Figure 250-3
(lb/ft^2)	4	4	9	See Figure 250-2	See Figure 250-3
Temperature (°C)	−20	−10	−1	+15	−10
(°F)	0	+15	+30	+60	+15

Courtesy of National Electrical Safety Code (NESC).
NOTE: Consult the 2007 NESC for detailed explanation of the new safety rules for overhead lines and to reference Figs. 250-2 and 250-3 prepared by the American Society of Civil Engineers (ASCE).

Sagging Conductors

Initial sag tables.   Used to determine the amount of sag to be placed on the conductors when they are installed.

Final sag tables.   Used in line design to determine clearances during various temperature conditions.

Ruling span = *average span* + 2/3 (*maximum span − average span*)

Span sag = [*span/(ruling span)*]^2 × *ruling span sag*

TABLE 2.10   Initial Sag Table for Various Span Lengths of 1/0 ACSR Conductor to Be Installed at 40% of Rated Ultimate Strength

	Distribution sag table Initial sag								
	Span length (ft)—Sag (in)				Conductor raven 1/0 (6/1) ACSR Maximum tension 40% ULT. = 1712 lb				
Temp.	100.00	110.00	120.00	130.00	140.00	150.00	160.00	170.00	180.00
−20.00	1.32	1.56	1.92	2.28	2.64	3.12	3.72	4.32	4.92
0.00	1.32	1.68	2.04	2.52	2.88	3.48	4.08	4.68	5.52
10.00	1.44	1.80	2.16	2.64	3.12	3.60	4.32	5.04	5.76
20.00	1.56	1.80	2.28	2.76	3.24	3.84	4.56	5.28	6.24
30.00	1.56	1.92	2.40	2.88	3.48	4.08	4.80	5.64	6.60
40.00	1.68	2.04	2.52	3.00	3.60	4.32	5.16	6.12	7.20
50.00	1.80	2.16	2.64	3.24	3.96	4.68	5.52	6.60	7.80
60.00	1.92	2.40	2.88	3.48	4.20	5.04	6.00	7.20	8.52
70.00	2.04	2.52	3.12	3.84	4.56	5.52	6.60	7.92	9.36
80.00	2.16	2.76	3.36	4.20	5.04	6.12	7.32	8.76	10.44
90.00	2.40	3.00	3.72	4.56	5.52	6.72	8.16	9.84	11.76
100.00	2.64	3.36	4.20	5.04	6.24	7.56	9.24	11.4	13.32
120.00	3.36	4.20	5.28	6.60	8.16	9.96	12.12	14.52	17.40
32.00	7.68	9.48	11.28	13.32	15.60	18.12	20.76	23.64	26.76
+1/2 ice 2-lb wind									
0.00	10.08	12.12	14.40	16.92	19.68	22.56	25.68	29.04	32.52
+1/2 ice 4-lb wind									
170.00	8.16	10.08	12.12	14.40	16.92	19.56	22.56	25.68	29.04

Courtesy of MidAmerican Energy Company.

TABLE 2.11   Final Sag Table for Various Span Lengths of 1/0 ACSR Conductor to Be Installed at 40% of Rated Ultimate Strength

Distribution sag table
Final sag

	Span length (ft)—Sag (in)				Conductor raven 1/0 (6/1) ACSR Maximum tension 40% ULT. = 1712 lb				
Temp.	100.00	110.00	120.00	130.00	140.00	150.00	160.00	170.00	180.00
0.00	1.44	1.80	2.16	2.64	3.24	3.84	4.68	5.52	6.60
10.00	1.56	1.92	2.40	2.88	3.48	4.20	5.04	6.12	7.32
20.00	1.68	2.16	2.64	3.24	3.84	4.68	5.64	6.84	8.16
30.00	1.92	2.28	2.88	3.60	4.32	5.28	6.36	7.68	9.36
40.00	2.04	2.64	3.24	3.96	4.92	6.00	7.20	8.88	10.68
50.00	2.28	2.88	3.60	4.56	5.52	6.84	8.40	10.20	12.48
60.00	2.64	3.36	4.20	5.16	6.48	7.92	9.72	11.88	14.52
70.00	3.00	3.84	4.92	6.12	7.56	9.36	11.52	14.04	16.92
80.00	3.60	4.56	5.76	7.32	9.00	11.16	13.56	16.44	19.44
90.00	4.32	5.52	7.08	8.76	10.92	13.20	15.96	18.60	21.48
100.00	5.28	6.84	8.52	10.20	12.24	14.40	16.80	19.56	22.44
120.00	6.36	7.80	9.48	11.52	13.56	15.96	18.60	21.48	24.60
32.00	8.64	10.56	12.60	14.88	17.40	20.04	23.04	26.16	29.52
+1/2 ice 2-lb wind									
0.00	10.08	12.12	14.40	16.92	19.68	22.56	25.68	29.04	32.52
+1/2 ice 4-lb wind									
170.00	8.88	10.80	12.96	15.36	17.88	20.64	23.76	26.88	30.36

Courtesy of MidAmerican Energy Company.

NOTES

NOTES

Chapter

3

Cable, Splices, and Terminations

15-kV Underground Distribution Cable Layers

Metallic conductor   The inner layer of a cable that carries the current.

Conductor strand shield layer   Applied between the conductor and the insulation and equalizes the charges around the conductor that are seeking the least-resistive path to ground.

Insulation   Compound that surrounds the conductor and isolates the metallic conductor from ground.

Semiconducting layer   Applied between the insulation and the neutral conductors or outer shield layer.

Outer layer metallic shield   Copper wires or metallic ribbon shield that provides a ground return in the event of short-circuit current due to cable failure.

Bedding tape   Layer installed between the insulation and the conductor sheath that allows for expansion within the layers and may or may not contain water-swelling resins or powders.

Outer jacket   Polyvinyl chloride (PVC) or polyethylene jacket that seals the cable from the environment.

Figure 3.1 Typical cable used for underground primary distribution. Aluminum-stranded conductor has a semiconducting extruded shielding, cross-linked polyethylene insulation, semiconducting extruded insulation shielding, surrounded by spiral-wound exposed tinned copper neutral conductor. The copper neutral conductor is often jacketed with polyethylene. (*Courtesy of Pirelli Cable Corporation.*)

Other Examples of Cable Layering

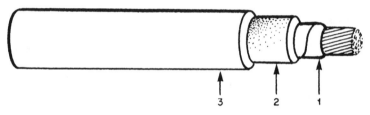

A

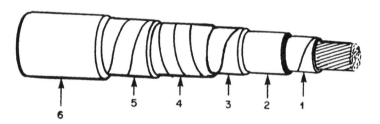

B

Figure 3.2 Cable construction. (A) Non-shielded cables: (1) strand shielding, (2) insulation, and (3) cable jacket. (B) Shielded power cable: (1) strand shielding, (2) insulation, (3) semiconducting material, (4) metallic shielding, (5) bedding tape, and (6) cable jacket.

Oil-filled cable

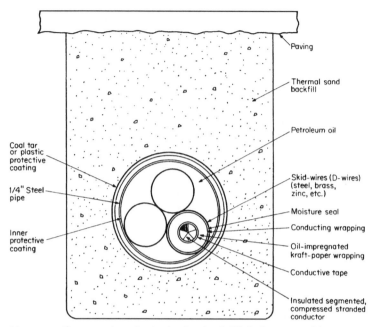

Paving

Thermal sand backfill

Petroleum oil

Coal tar or plastic protective coating

1/4" Steel pipe

Inner protective coating

Skid-wires (D-wires) (steel, brass, zinc, etc.)

Moisture seal

Conducting wrapping

Oil-impregnated kraft-paper wrapping

Conductive tape

Insulated segmented, compressed stranded conductor

Figure 3.3   Cross section of a single-circuit oil-filled pipe-type cable system.

Transmission cable

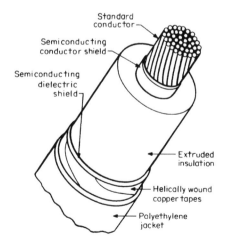

Standard conductor

Semiconducting conductor shield

Semiconducting dielectric shield

Extruded insulation

Helically wound copper tapes

Polyethylene jacket

Figure 3.4   Construction of solid-dielectric extruded-insulation, high-voltage transmission cable.

Self-contained transmission cable

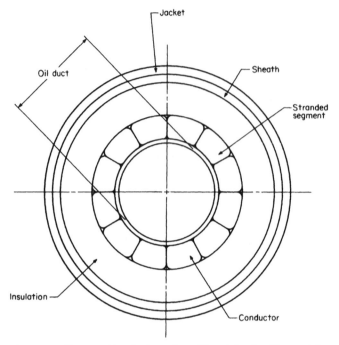

Figure 3.5   Cross-sectional view of a self-contained cable used for underground transmission circuits.

Conductor Metal Types

- Aluminum
- Copper

Conductor Composition

- Stranded
- Solid

Conductor Shapes

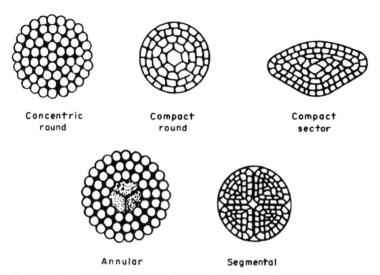

| Concentric round | Compact round | Compact sector |

| Annular | Segmental |

Figure 3.6   Diagrams of common shapes of stranded conductors.

Insulation Compounds

EPR   Ethylene propylene rubber

XLPE   Cross-linked polyethylene

TRXLPE   Tree-retardant polyethylene

HMWPE   High-molecular-weight polyethylene

PILC   Paper-insulated lead-covered

Insulation Thickness

Distribution voltage	100% Insulation level	133% Insulation level
5 kV	90 mil	90 mil
15 kV	175 mil	220 mil
25 kV	260 mil	345 mil
35 kV	345 mil	420 mil

100% level   Ground fault will be cleared within 1 minute.

133% level   Fault-clearing requirements will not meet 100% criteria but fault will clear within 1 hour.

Sample Cable Specifications

15-kV XLP insulated 4/0-AWG aluminum cable with jacketed concentric neutral

This specification covers a typical 15-kV, cross-linked polyethylene (XLP) insulated, shielded, concentric neutral cable with an overall insulating jacket. Application of this cable is for 60-Hz Y-connected system with a grounded neutral operating at 13,200Y/7620 volts.

- The cable shall be suitable for installing in ducts or direct burial in the earth in wet or dry locations.

- The cable shall be furnished with a 4/0-AWG, 19-strand aluminum conductor.

- The maximum conductor temperature shall be 90°C normal, 130°C emergency, and 250°C short-circuit.

Conductor.   Alloy 1350 aluminum. The stranding shall be class-B compressed stranded, per ASTM B231.

Conductor strand shield layer.   The conductor strand shield layer shall be extruded over the stranded conductor with good concentricity and shall be a black insulating material compatible with the conductor and the primary insulation. The strand shield layer shall have a specific inductive capacitance of 10 or greater. In addition to its energy-suppression function, this layer shall provide stress relief at the interface with the primary insulation. The strand shield layer shall be easily removed from the conductor with the use of commonly available tools and shall not have a thickness at any part of the cable of less than 12 mils when measured over the top of the strands. The outer surface of the extruded strand shield layer shall be cylindrical, smooth, and free of significant irregularities.

Primary insulation.   The primary insulation shall be 175 mils of extruded, tree-retardant, cross-linked polyethylene compound. The primary insulation shall be extruded directly over the conductor strand shield layer with good concentricity and shall be compatible with the conductor energy-suppression layer. The outer surface of the extruded primary insulation shall be cylindrical, smooth, and free of significant irregularities.

Insulation shielding system.   The cable insulation shielding system shall consist of semiconducting, extruded, strippable, cross-linked material. The insulation shield minimum point, maximum point, and maximum indent thickness shall be 0.800 to 0.900 inch.

Concentric neutral.   A concentric neutral consisting of 16- #14 bare copper wires, in accordance with ANSI B-3, shall be spirally wound over the insulation shield, with uniform spacing between wires and a lay of not less than 6 not more than 10 times the cable diameter.

Jacket.   The jacket shall be a linear low-density polyethylene extruded over and encapsulating the concentric neutral. The minimum average thickness shall be 50 mils and the minimum thickness at any point on the cable shall not be less than 80 percent of the minimum average.

Identification.   The outer surface of each jacketed cable shall be durably marked by sharply defined surface printing. The surface marking shall be as follows:

- Manufacturer's name
- Conductor size
- Conductor material
- Rated voltage
- Year of manufacture
- Symbol for supply cable per NESC C2-350.G

Cable-diameter dimensional control.   The overall diameter of the 15-kV insulated cable shall not exceed 1.35 inches.

Production tests

Conductor tests.   The conductor shall be tested in accordance with AEIC CS95-94 and ICEA Standard S-66-524.

Conductor strand shield layer.   The integrity of the conductor strand shield layer shall be continuously monitored prior to or during the extrusion of the primary insulation using an electrode system completely covering the full circumference energy-suppression layer at a test voltage of 2 kV direct-current.

Insulation tests

Preliminary tank test.   Each individual length of insulated conductor, prior to the application of any further coverings, shall be immersed in a water tank for a minimum period of 24 hours. At the end of 16 hours (minimum), each length of insulated conductor shall pass a DC voltage withstand test applied for 5 minutes, the applied DC voltage should be 70 kV.

At the end of 24 hours (minimum), each length of insulated conductor shall pass an AC voltage withstand test applied for 5 minutes, the applied AC voltage should be 35 kV. Each length of insulated conductor following the 5-minute AC voltage test shall have an insulation resistance, corrected to 15.6° C (60° F), of not less than the value of R, as calculated from the following formula (after electrification for 1 minute):

$$R = K \log_{10} D/d$$

R = insulation resistance in MΩ/1000 ft
K = 21,000 MΩ/1000 ft
D = diameter over insulation
d = conductor diameter

Final test on shipping reel.   Each length of completed cable shall pass an AC voltage withstand test applied for 1 minute, the applied AC voltage should be 35 kV. Each length of completed cable, following the 1-minute AC voltage test, shall have an insulation resistance not less than that specified above.

Dimensional verification

- Each length of insulated conductor shall be examined to verify that the conductor is the correct size.
- Each length of insulated conductor shall be measured to verify that the insulation wall thickness is of the correct dimension.
- Each length of completed cable shall be measured to verify that the jacket wall thickness is of the correct dimension.

Conductor and shield continuity.   Each length of completed cable shall be tested for conductor and shield continuity.

Test reports.   Certified test reports of all cable will be required, one copy of test reports to be sent to the ordering party.

Shipping lengths.   The shipping lengths will be specified on the inquiry and purchase order. Unless otherwise specified on the purchase order, the cable shall be furnished in one continuous reel.

Reels.   Reels shall be heavy-duty wooden reels (nonreturnable, if possible) properly inspected and cleaned to ensure that no nails, splinters, etc. may be present and possibly damage the cable. Each end of the cable shall be firmly secured to the reel with nonmetallic rope. Wire strand is not acceptable for securing cable to the reel. Deckboard or plywood covering secured with steel banding is required over the cable.

Cable sealing.   Each end of the cable shall be sealed hermetically before shipment using either heat-shrinkable polyethylene caps or hand taping, whichever the manufacturer recommends. Metal clamps that may damage pulling equipment are not acceptable. The end seals must be adequate to withstand the rigors of pulling in.

Preparation for transit.   Reels must be securely blocked in position so that they will not shift during transit. The manufacturer shall assume responsibility for any damage incurred during transit that is the result of improper preparation for shipment.

Reel marking.   Reel markings shall be permanently stenciled on one flange of the wooden reel. Such markings shall consist of the purchaser's requirements as indicated elsewhere in this specification, but shall consist, as a minimum, of a description of the cable, purchaser's order number, manufacturer's order number, date of shipment, footage, and reel number. An arrow indicating direction in which reel should be rolled shall be shown.
(End of Sample cable specification.)

Electrical stress distribution

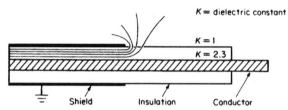

Figure 3.7   Electric stress distribution for high-voltage solid-dielectric insulated shielded cable with the shield and insulation partially removed from cable conductor. Note high concentration of electrical stress at the edge of the cable shield.

Terminations (electrical stress minimization)

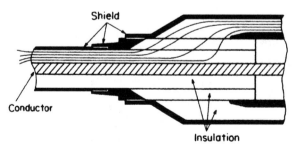

Figure 3.8   Electric stress distribution in a completed prefabricated splice installed on a solid-dielectric-insulated shielded high-voltage cable. Stress in transition area from cable to splice is evenly distributed.

Live-front applications

Riser pole or lugged terminations

Correct installation of termination

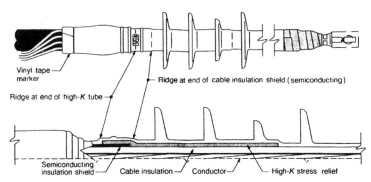

Vinyl tape marker

Ridge at end of high-*K* tube

Ridge at end of cable insulation shield (semiconducting)

Semiconducting insulation shield

Cable insulation

Conductor

High-*K* stress relief

Figure 3.9 Typical termination make up for use with single-phase 7620-volt cable used for underground distribution. (*Courtesy of 3M Co.*)

Dead-front applications

Load-break elbow termination

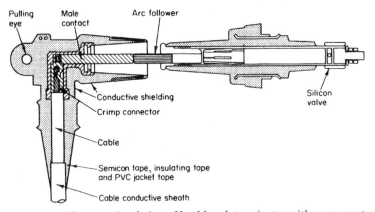

Pulling eye

Male contact

Arc follower

Conductive shielding

Crimp connector

Silicon valve

Cable

Semicon tape, insulating tape and PVC jacket tape

Cable conductive sheath

Figure 3.10 Cross-sectional view of load-break terminator with component parts identified.

Dead-front load-break elbow termination detail

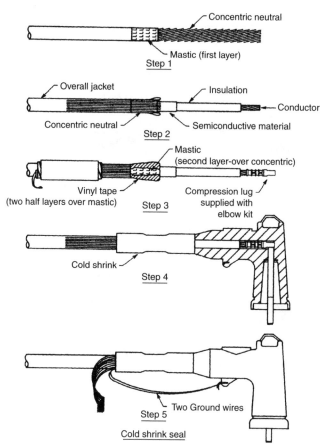

Figure 3.11   The steps involved in the make-up of a cold-shrink seal of a dead-front termination. For proper dimensioning of the cable layers, consult the instructions enclosed in the manufacturer's elbow installation kit. (*Courtesy of MidAmerican Energy Co.*)

In-line cable splices

Cold-shrink splices

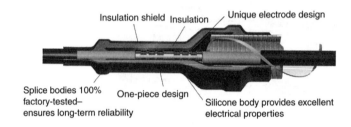

Insulation shield   Insulation   Unique electrode design

Splice bodies 100%   One-piece design
factory-tested–
ensures long-term reliability   Silicone body provides excellent
                                         electrical properties

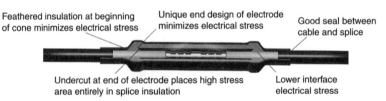

Feathered insulation at beginning   Unique end design of electrode
of cone minimizes electrical stress   minimizes electrical stress   Good seal between
                                                                          cable and splice

Undercut at end of electrode places high stress   Lower interface
area entirely in splice insulation   electrical stress

Figure 3.12   Cross-sectional view of a prefabricated cold shrink splice installed on concentric neutral cross-linked polyethylene-insulated underground cable. (*Courtesy of 3M Co.*)

TABLE 3.1   Cable Ampacity Tables

Material, application, & size	Normal conditions				Emergency conditions				Assumptions
	Max amps	Maximum MVA at			Max amps	Maximum MVA at			
		12.47	13.2	13.8		12.47	13.2	13.8	
Aluminum									
Single-phase application									Earth Interface Temperature — 60°C
2	165	3.6	3.8	3.9	190	4.1	4.3	4.5	Normal Cable Temperature — 90°C
1	190	4.1	4.3	4.5	220	4.8	5.0	5.3	Emergency Cable Temperature — 130°C
1/0	215	4.6	4.9	5.1	245	5.3	5.6	5.9	Ambient Earth Temperature — 20°C
2/0	250	5.4	5.7	6.0	290	6.3	6.6	6.9	Earth Resistivity — 90°C-cm/W
3/0	280	6.0	6.4	6.7	320	6.9	7.3	7.6	Depth of Burial — 36″
4/0	320	6.9	7.3	7.6	370	8.0	8.5	8.8	Daily Load Factor (Avg/Peak) — 80%
Three-phase application (triplexed arrangement)									Three-Phase Load — Balanced
2	155	3.3	3.5	3.7	180	3.9	4.1	4.3	Single-Phase Application — Full Neutral (Concentric)
1	175	3.8	4.0	4.2	200	4.3	4.6	4.8	Three-Phase Application — Reduced Neutral (Concentric)
1/0	200	4.3	4.6	4.8	230	5.0	5.3	5.5	
2/0	230	5.0	5.3	5.5	265	5.7	6.1	6.3	Single-Conductor Cables with XLPE, TR-XLPE, or EPR Insulation
3/0	260	5.6	5.9	6.2	300	6.5	6.9	7.2	
4/0	290	6.3	6.6	6.9	335	7.2	7.7	8.0	
250	320	6.9	7.3	7.6	370	8.0	8.5	8.8	
350	385	8.3	8.8	9.2	445	9.6	10.2	10.6	
400	415	9.0	9.5	9.9	475	10.3	10.9	11.4	
500	460	9.9	10.5	11.0	530	11.4	12.1	12.7	
600	495	10.7	11.3	11.8	565	12.2	12.9	13.5	
750	550	11.9	12.6	13.1	635	13.7	14.5	15.2	
1000	620	13.4	14.2	14.8	715	15.4	16.3	17.1	

(Continued)

73

TABLE 3.1 Cable Ampacity Tables (Continued)

Material, application, & size	Normal conditions				Emergency conditions			
	Max amps	Maximum MVA at			Max amps	Maximum MVA at		
		12.47	13.2	13.8		12.47	13.2	13.8
Copper								
Single-phase application								
2	215	4.6	4.9	5.1	250	5.4	5.7	6.0
1	245	5.3	5.6	5.9	280	6.0	6.4	6.7
1/0	275	5.9	6.3	6.6	315	6.8	7.2	7.5
2/0	315	6.8	7.2	7.5	360	7.8	8.2	8.6
3/0	355	7.7	8.1	8.5	410	8.9	9.4	9.8
4/0	410	8.9	9.4	9.8	470	10.2	10.7	11.2
Three-phase application								
2	200	4.3	4.6	4.8	230	5.0	5.3	5.5
1	225	4.9	5.1	5.4	260	5.6	5.9	6.2
1/0	260	5.6	5.9	6.2	300	6.5	6.9	7.2
2/0	290	6.3	6.6	6.9	335	7.2	7.7	8.0
3/0	330	7.1	7.5	7.9	380	8.2	8.7	9.1
4/0	370	8.0	8.5	8.8	425	9.2	9.7	10.2
250	405	8.7	9.3	9.7	465	10.0	10.6	11.1
350	480	10.4	11.0	11.5	550	11.9	12.6	13.1
400	510	11.0	11.7	12.2	600	13.0	13.7	14.3
500	550	11.9	12.6	13.1	635	13.7	14.5	15.2
600	580	12.5	13.3	13.9	665	14.4	15.2	15.9
750	630	13.6	14.4	15.1	725	15.7	16.6	17.3
1000	700	15.1	16.0	16.7	805	17.4	18.4	19.2

Courtesy of MidAmerican Energy Company.

A multiplication factor of 1.15 is used to go from normal to emergency conditions. This approximately accounts for the effect of temperature on the conductor material and other cable components including dielectric losses. In accordance with AEIC standards, the 130°C operating temperature should be limited to 1500 hours for the life of the cable and 100 hours in a 12-month period.

Temperature Correction Tables

TABLE 3.2 Ampacity Values as the Cable Temperature Is Varied

Cable temp. °C	Adjustment factor	Cable									
		1/0 AL set: direct		4/0 AL set: direct		750 AL set: direct		750 AL set: duct	
		Amps	% Change	Amps	% Change	Amps	% Change	Amps	% Change
75	0.908	182		263		499		422	
80	0.941	188	3.6%	273	3.6%	517	3.6%	437	3.6%
85	0.971	194	3.2%	282	3.2%	534	3.2%	452	3.2%
90	**1.000**	**200**	3.0%	**290**	3.0%	**550**	3.0%	**465**	3.0%
95	1.027	205	2.7%	298	2.7%	565	2.7%	478	2.7%
100	1.053	211	2.5%	305	2.5%	579	2.5%	489	2.5%
105	1.077	215	2.3%	312	2.3%	592	2.3%	501	2.3%
110	1.100	220	2.1%	319	2.1%	605	2.1%	511	2.1%
115	1.122	224	2.0%	325	2.0%	617	2.0%	522	2.0%
120	1.143	229	1.9%	331	1.9%	628	1.9%	531	1.9%
125	1.162	232	1.7%	337	1.7%	639	1.7%	541	1.7%
130	1.181	236	1.6%	343	1.6%	650	1.6%	549	1.6%
135	1.200	240	1.5%	348	1.5%	660	1.5%	558	1.5%
140	1.217	243	1.5%	353	1.5%	669	1.5%	566	1.5%
145	1.234	247	1.4%	358	1.4%	679	1.4%	574	1.4%
150	1.250	250	1.3%	362	1.3%	687	1.3%	581	1.3%

Courtesy of MidAmerican Energy Company.
Ambient Earth temperature   = 20°C
Earth Thermal Resistivity, RHO = 90°C-cm/W
Load Factor   = 100%
Depth of Burial   = 36" for direct buried, 30" to top of duct

TABLE 3.3 Multiplication Factors as the Aluminum Cable Temperature and Ambient Earth Temperatures Are Varied from a Standard Cable Temperature of 90°C and an Ambient Earth Standard of 20°C

Aluminum		Ambient earth temperature, T_a,°C						
		0	5	10	15	20	25	30
	75	1.060	1.024	0.987	0.948	0.908	0.866	0.821
	80	1.086	1.052	1.016	0.979	0.941	0.901	0.859
	85	1.111	1.078	1.043	1.008	0.971	0.933	0.893
	90	1.134	1.102	1.069	1.035	1.000	0.964	0.926
	95	1.156	1.125	1.093	1.061	1.027	0.992	0.956
Cable	100	1.177	1.147	1.116	1.085	1.053	1.019	0.985
temperature	105	1.197	1.168	1.138	1.108	1.077	1.045	1.012
T_C	110	1.216	1.188	1.159	1.130	1.100	1.069	1.037
°C	115	1.234	1.207	1.179	1.151	1.122	1.092	1.061
	120	1.252	1.225	1.198	1.171	1.143	1.114	1.084
	125	1.268	1.243	1.217	1.190	1.162	1.134	1.106
	130	1.284	1.259	1.234	1.208	1.181	1.154	1.126
	135	1.300	1.276	1.251	1.225	1.200	1.173	1.146
	140	1.315	1.291	1.267	1.242	1.217	1.192	1.165
	145	1.329	1.306	1.282	1.258	1.234	1.209	1.184
	150	1.343	1.320	1.297	1.274	1.250	1.226	1.201

Courtesy of MidAmerican Energy Company.

TABLE 3.4 Multiplication Factors as the Copper Cable Temperature and Ambient Earth Temperatures Are Varied from a Standard Cable Temperature of 90°C and an Ambient Earth Standard of 20°C

Copper		Ambient earth temperature, T_a,°C						
		0	5	10	15	20	25	30
	75	1.060	1.024	0.987	0.948	0.908	0.865	0.821
	80	1.086	1.051	1.016	0.979	0.940	0.900	0.858
	85	1.111	1.077	1.043	1.008	0.971	0.933	0.893
	90	1.134	1.102	1.069	1.035	1.000	0.964	0.926
	95	1.156	1.125	1.094	1.061	1.027	0.992	0.956
Cable	100	1.177	1.147	1.117	1.085	1.053	1.020	0.985
temperature	105	1.197	1.169	1.139	1.109	1.077	1.045	1.012
T_C	110	1.217	1.189	1.160	1.131	1.100	1.069	1.038
°C	115	1.235	1.208	1.180	1.152	1.123	1.093	1.062
	120	1.253	1.226	1.199	1.172	1.144	1.115	1.085
	125	1.270	1.244	1.218	1.191	1.164	1.136	1.107
	130	1.286	1.261	1.235	1.209	1.183	1.156	1.128
	135	1.301	1.277	1.252	1.227	1.201	1.175	1.148
	140	1.316	1.293	1.269	1.244	1.219	1.193	1.167
	145	1.331	1.308	1.284	1.260	1.236	1.211	1.185
	150	1.345	1.322	1.299	1.276	1.252	1.228	1.203

Courtesy of MidAmerican Energy Company.

Temperature Correction Formula (Information Courtesy of Southwire Company)

Adjustment for other temperatures

It is often necessary to determine the ampacity at conditions other than those specified on published tables. Any value of ampacity may be adjusted for a change in one or more basic parameters by using the following equations:

Copper conductors

$$I' = I\sqrt{\frac{T'_C - T'_A}{T_C - T_A} \cdot \frac{234 + T_C}{234 + T'_C}} \quad \text{amps} \qquad (3.1)$$

Aluminum conductors

$$I' = I\sqrt{\frac{T'_C - T'_A}{T_C - T_A} \cdot \frac{228 + T_C}{228 + T'_C}} \quad \text{amps} \qquad (3.2)$$

where the primed (') values are the revised parameters.

Sample calculation

A 90°C rated copper cable has a published ampacity of 500 amps under a given set of conditions which includes an ambient temperature (T_A) of 40°C. Find the ampacity at a conductor temperature of 80°C and ambient temperature of 50°C.

Using Eq. (3.1)

$$I' = I\sqrt{\frac{T'_C - T'_A}{T_C - T_A} \cdot \frac{234 + T_C}{234 + T'_C}} \quad \text{amps}$$

$$I' = 500\sqrt{\frac{80 - 50}{90 - 40} \cdot \frac{234 + 90}{234 + 80}} \quad \text{amps}$$

$$I' = (500) \cdot (0.787) = 393 \quad \text{amps}$$

NOTES

Chapter

4

Distribution Voltage Transformers

Distribution Voltage Transformer Definition

A distribution transformer is a static device constructed with two or more windings used to transfer alternating-current electric power by electromagnetic induction from one circuit to another at the same frequency, but with different values of voltage and current. The purpose of a distribution voltage transformer is to reduce the primary voltage of the electric distribution system to the utilization voltage. The product of the primary voltage and the primary current are equal to the product of the secondary voltage and the secondary current.

Alternating-Current Electric Voltage Step-Down Transformer

Figure 4.1 shows the elementary parts of an alternating-current electric voltage step-down transformer.

Transformer Components

Coil   Insulated wire that is continuously wrapped around the transformer's core.
Core   Thin metal sheets that are arranged to form a closed magnetic circuit.
Primary bushing   The high-voltage attachment point on the transformer.
Secondary bushing   The low-voltage attachment point on the transformer.
Primary winding   The coil to which the input voltage is applied.

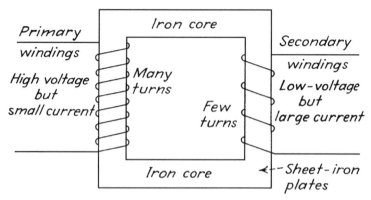

Figure 4.1 The secondary voltage is based upon the primary voltage, the number of turns of the primary windings, and the number of turns of the secondary windings.

Secondary winding The coil from which the output voltage is derived.
Step-down transformer The input voltage of the transformer is more than the output voltage.
Step-up transformer The input voltage of the transformer is less than the output voltage.

Transformer Formulas

Voltage, V = Volts per turn × Number of turns, N
Voltage, primary winding = V_p
Voltage, secondary winding = V_s
Current, primary winding = I_p
Current, secondary winding = I_s
Number of turns in the primary winding = N_p
Number of turns in the secondary winding = N_s

Transformer turns ratio formula

$$V_p/N_p = V_s/N_s$$

or

$$V_s = (N_s/N_p) \times V_p$$

Example: If a distribution transformer has a primary winding of 2540 turns and a secondary winding of 40 turns and the primary winding voltage equals 7620 volts, determine the secondary winding voltage.

Solution: $V_s = \dfrac{N_s}{N_p} \times V_p$

$$V_s = \dfrac{40}{2540} \times 7620$$

$$V_s = 120 \text{ volts}$$

Primary to secondary voltage and current relationship

$$V_p \times I_p = V_s \times I_s$$

or

$$I_p = \dfrac{V_s}{V_p} \times I_s$$

Example: If the primary winding voltage of a transformer is 7620 volts, the secondary winding voltage is 120 volts, and the secondary current, as a result of the load impedance, is 10 amps, then determine the primary current.

Solution: $I_p = \dfrac{V_s}{V_p} \times I_s$

$$I_p = \dfrac{120 \times 10}{7620} = 0.16 \text{ amp}$$

Distribution Transformer Ratings

For overhead and pad-mounted distribution transformer ratings, see Table 4.1.

TABLE 4.1   Distribution Transformer Ratings (kVA)

Overhead type		Pad-mounted type	
Single phase	Three phase	Single phase	Three phase
5	15	25	75
10	30	37½	112½
15	45	50	150
25	75	75	225
37½	112½	100	300
50	150	167	500
75	225		750
100	300		1000
167	500		1500
250			2000
333			2500
500			

Distribution Transformer Specifications

Overhead conventional distribution transformer

A typical specification for conventional single-phase overhead distribution transformers (Fig. 4.2) would read as follows:

> This standard covers the electrical characteristics and mechanical features of single-phase, 60-Hz, mineral-oil immersed, self-cooled, conventional, overhead-type distribution transformers rated: 15, 25, 50, 75, 100, and 167 kVA; with high voltage: 7200/12,470Y, 7620/13,200Y, or 7970/13,800Y volts; and low voltage: 120/240 volts; with no taps. The unit shall be of "conventional" transformer design. There shall be no integral fuse, secondary breaker, or overvoltage protection.
>
> All requirements shall be in accordance with the latest revision of ANSI C57.12.00 and ANSI C57.12.20, except where modified by this standard.
>
> The transformer shall be bar coded in accordance with the proposed EEI *Bar Coding Distribution Transformers: A Proposed EEI Guideline*. The manufacturer and serial number shall be bar coded on the nameplate. The serial number, manufacturer, and the Company Item ID (if given) shall be bar coded on a temporary label.
>
> The transformer shall meet interchangeability dimensions for single-position mounting, as shown in Figures 1 and 9 of ANSI C57.12.20.
>
> The two primary bushings shall be wet-process porcelain cover mounted. The primary connectors shall be tin-plated bronze of the eye-bolt type. The H1 bushing is to be supplied with a wildlife protector and an eyebolt connector equipped with a covered handknob tightening device. The eyebolt connector shall accept a 5/16-inch diameter knurled stud terminal with the

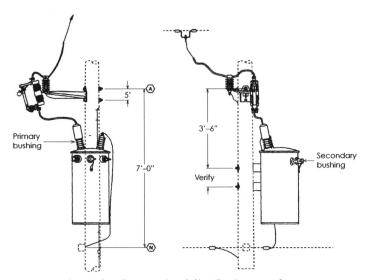

Figure 4.2 An overhead conventional distribution transformer.

wildlife protector in place. The secondary bushings shall be wet-process porcelain or epoxy, mounted on the tank wall. Secondary connectors shall be per ANSI C57.12.20, section 6.1.2, except that terminals on 75-kVA 120/240-volt units shall accommodate up to 1000 kcmil-61 stranded conductor.

The cover of the transformer shall be coated for wildlife protection. The coating shall have a rating of at least 10 kV AC rms. The insulating finish shall be capable of withstanding 10 kV at a 200-volt per second rate of rise, tested per ASTM D149-87, using electrode type 3 (1/4-inch cylindrical rods).

The transformer tank and cover shall be designed to withstand an internal pressure of 20 psig without tank rupture or bushing or cover displacement, and shall withstand an internal pressure of 7 psig without permanent distortion. The transformer tank shall retain its shape when the cover is removed.

Support lugs for single-position mounting shall be provided and shall be designed to provide a static-load safety factor of five. Lifting lugs shall be provided, with location and safety factor, as specified in ANSI C57.12.20. The tank and cover finish shall be light gray No. 70 (Munsell 5BG7.0/0.4). The interior shall be coated with a durable protective finish from the top to at least below the top liquid level.

The transformer shall be equipped with a tank ground per ANSI C57.12.20, paragraph 6.5.4.1 with eyebolt to accept an AWG 6 solid copper ground wire, a tank ground for ground strap connection, and a ground strap, properly sized to carry load and fault current, connected between the secondary neutral terminal and the tank ground. The grounding strap shall be supplied with a connection hole at each end. "J" hook or slotted connection points shall not be allowed.

The transformer shall be equipped with neoprene or Buna-N gaskets on all primary and secondary bushings and tank fittings.

A Qualitrol 202-032-01 automatic pressure relief device, with operation indication, or approved equivalent; or a cover assembly designed for relief of excess pressure in accordance with ANSI C57.12.20, section 6.2.7.2, shall be supplied on the transformer.

The final core and coil assembly shall result in a rigid assembly that maintains full mechanical, electrical, and dimensional stability under fault conditions.

The coil-insulation system shall consist of electrical kraft paper layer insulation coated on both sides with thermosetting adhesive. Curing of the coils shall cause a permanent bond, layer to layer and turn to turn.

The internal primary leads shall have sufficient length to permit cover removal and lead disconnection from the high-voltage bushings. These leads shall be adequately anchored to the coil assembly to minimize the risk of damaging the coil connection during cover removal.

The internal secondary leads shall be identified by appropriate markings, embossed or stamped on a visible portion of the secondary lead near the end connected to the bushing. These leads shall have sufficient length to permit series or parallel connection of the secondary windings. If the internal secondary leads are soft aluminum, a hard aluminum-alloy tab shall be welded to the bushing end of the lead for connection to the bushing.

The transformer should be designed so that the voltage regulation shall not exceed 2.5% at full load and 0.8 power factor for all transformers covered in this specification.

All transformers shall be filled with Type I insulating oil that meets the following requirements:

- All applicable requirements of ASTM D 3487, paragraph 82a.
- Minimum breakdown value of 28 kV per ASTM D 1816, paragraph 84a using a 0.4-inch gap.
- The PCB content shall be less than 2 parts per million, which shall be noted upon the nameplate of each unit. In addition, the manufacturer shall provide and install a "No PCB" identification on the tank wall, near the bottom of the tank, centered below the low-voltage bushings. The decal shall be not less than 1 inch by 2 inch in size and shall indicate "No PCBs" in blue lettering on a white background, and be sufficiently durable to remain legible for the life of the transformer.

Overhead Completely Self-Protected (CSP) Distribution Transformer

A typical specification for a completely self-protected single-phase overhead distribution transformers (Fig. 4.3) would read as follows:

Distribution transformers; overhead type; single-phase; 25-kVA capacity; 13,200 GrdY/7620-120/240 volts; 95-kV BIL; self-protected, with primary weak-link fuse and secondary circuit breaker; red overload warning light; oil-insulated, self-cooled (OISC); no taps; 60 Hz; one surge arrester; one primary bushing, with wildlife protector and eyebolt connector equipped with an insulated hand knob-tightening device designed to accept a 5/16-inch diameter knurled stud terminal with wildlife protector in place; tank ground, with eyebolt designed to accept No. 6 solid copper ground wire; a tank ground terminal for a ground strap connection; a ground strap, properly sized to carry load and fault current, connected between the secondary neutral terminal and the tank ground terminal; cover coated to provide 10-kVAC rms protection from wildlife; 65°C rise rating, sky gray in color, Qualitrol No. 202-032-01 Automatic Pressure Relief Device, with operation indication, or approved equivalent; and standard accessories.

The primary weak-link fuse, secondary circuit breaker, and red overload warning light shall be coordinated and compatible with the nameplate ratings.

The arrester shall be a 9/10-kV, direct-connected lead-type distribution-class surge arrester (in accordance with the latest revision of ANSI standard C62.2), with ground lead disconnector, and wildlife protector. The arrester primary lead shall have a minimum length of 24 inch with a weatherproof jacket, for connection to the line side of a current-limiting fuse to be installed on the primary bushing in the field.

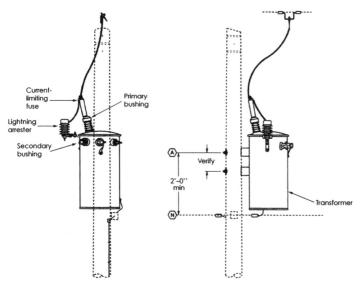

Figure 4.3 An overhead CSP distribution transformer.

All requirements shall be in accordance with the latest revision of ANSI standard C57.12.20, except where modified by this specification.

The transformer shall be bar coded in accordance with the proposed EEI *Bar Coding Distribution Transformers: A Proposed EEI Guideline*. The manufacturer and serial number shall be bar coded on the nameplate. The serial number, manufacturer, and the company item ID (if given) shall be bar coded on a temporary label.

The transformer shall meet interchangeability dimensions for single position mounting, as shown in Figures 1 and 9 of ANSI C57.12.20.

All transformers shall be filled with Type I insulating oil that meets the following requirements:

- All applicable requirements of ASTM D 3487, paragraph 82a.
- Minimum breakdown value of 28 kV per ASTM D 1816, paragraph 84a using a 0.4-inch gap.
- The PCB content shall be less than 2 parts per million, which shall be noted upon the nameplate of each unit. In addition, the manufacturer shall provide and install a "No PCB" identification on the tank wall, near the bottom of the tank, centered below the low-voltage bushings. The decal shall be not less than 1 inch by 2 inch in size and shall indicate "No PCBs" in blue lettering on a white background, and be sufficiently durable to remain legible for the life of the transformer.

This specification can be changed to meet the kVA capacity and voltages required. Changes in the high voltage stated in this specification would require changes in the related items in the specification to obtain the proper insulation and surge-withstand capabilities.

Single-Phase Pad-Mounted Distribution Transformer

A typical specification for a single-phase pad-mounted distribution transformer (Fig. 4.4) would read as follows:

Distribution transformers; low profile; dead front; pad mounted; single phase; 50-kVA capacity; 13,200 GrdY/7620-120/240 volts; OISC; 95-kV BIL; 60 Hz; 65°C rise rating; and no taps on the primary winding.

All requirements shall be in accordance with the latest revisions of ANSI C57.12.25, except where modified by this specification.

The transformer shall be equipped with two primary bolted-on bushing wells, rated 8.3 kV and 200 A, in accordance with the latest revision of ANSI/IEEE 386. The bushing wells shall be equipped with 8.3/14.4-kV

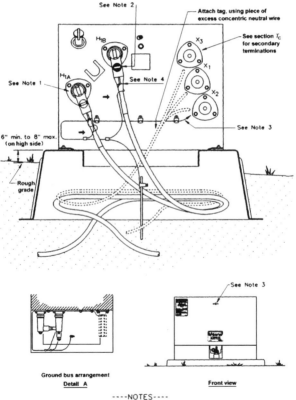

----NOTES----

1. Attach a drain wire (a piece of excess concentric neutral wire) from each bushing insert to the bushing well clamping flange.

2. Install fault indicator sensor on the source side elbow. See drawing FI 010 for fault indicator installation instructions.

3. Attach end of ground bus to the neutral secondary terminator. See detail "A."

4. One concentric neutral strand is to be connected to the drain wire lobe

Figure 4.4 A single-phase pad-mounted distribution transformer.

200-A load break bushing inserts to accommodate load break elbows for sectionalizing cable in a loop-feed primary system.

The secondary terminals are to be equipped with bolted standard copper-threaded studs only.

A ground pad shall be provided on the outer surface of the tank. The low-voltage neutral shall have a fully insulated bushing. A removable ground strap sized for the rating of the transformer shall be provided and connected between the neutral bushing and a ground terminal on the tank of the transformer.

The high-voltage and low-voltage ground pads inside the transformer compartment shall be equipped with eyebolt connectors to accept 1/0 bare copper wire.

The transformer shall be equipped with an RTE Bay-O-Net, or approved equivalent, externally replaceable overload sensing primary fuse for coordinated overload and secondary fault protection coordinated and in series with an 8.3-kV partial-range current-limiting fuse. The current-limiting fuse shall be sized so that it will not operate on secondary short circuits or transformer overloads. The current-limiting fuse shall be installed under oil inside of the transformer tank. Each load-sensing and current-limiting fuse device shall be capable of withstanding a 15-kV system-recovery voltage across an open fuse.

The transformer shall be equipped with an oil drip shield directly below the Bay-O-Net fuse holder to prevent oil from dripping on the primary elbows or bushings during removal of the fuse.

The transformer housing shall be designed to be tamper-resistant to prevent unauthorized access.

The transformer shall be bar coded in accordance with the proposed EEI *Bar Coding Distribution Transformers: A Proposed EEI Guideline.* The manufacturer and serial number shall be bar coded on the nameplate. The serial number, manufacturer, and the company item ID (if given above) shall be bar coded on a temporary label. Specifications about transformers that need to be filled with Type-I insulating oil have been discussed earlier in this chapter.

All transformers shall be filled with Type I insulating oil that meets the following requirements:

a. All applicable requirements of ASTM D 3487, paragraph 82a.
b. Minimum breakdown value of 28 kV per ASTM D 1816, paragraph 84a using a 0.4-inch gap.
c. The PCB content shall be less than 2 parts per million, which shall be noted upon the nameplate of each unit. In addition, the manufacturer shall provide and install a "No PCB" identification on the tank wall, near the bottom of the tank, centered below the low-voltage bushings. The decal shall be not less than 1 inch by 2 inch in size and shall indicate "No PCBs" in blue lettering on a white background, and be sufficiently durable to remain legible for the life of the transformer.

This specification can be changed to meet the kVA capacity and voltages required. Changes in the high voltage stated in this specification would require changes in the related items in the specification to obtain the proper insulation and surge-withstand capabilities.

Three-Phase Pad-Mounted Distribution Transformer

A typical specification for a three-phase pad-mounted distribution transformer (Fig. 4.5) would read as follows:

Distribution transformers; three-phase; deadfront; 300-kVA capacity; high-voltage windings rated 13,800 GrdY/7970 volts; low-voltage windings rated 208 GrdY/120 volts; the high-voltage winding shall have four 2 1/2 percent taps below rated voltage; and the no-load tap-changer switch is to be located in the high-voltage compartment.

All requirements shall be in accordance with the latest revision of ANSI C57.12.26, except where changed by this specification.

The transformer housing shall be designed to be tamper-resistant to prevent unauthorized access in accordance with the latest revision to the

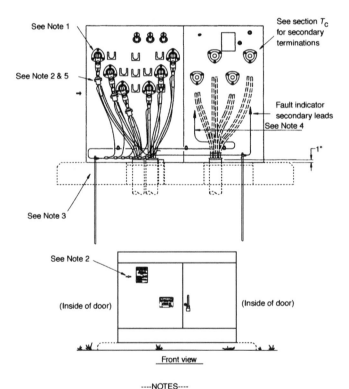

----NOTES----

1. Attach drain wire from each bushing insert to the bushing well clamping flange.
2. Install fault indicator current sensors on the source side elbows between concentric neutral strands and elbow. See drawing FI 020 for fault indicator installation instructions.
3. Concrete pad, conduit elbows and ground rod sleeve to be furnished by customer.
4. Attach end of ground bus to the neutral secondary terminator.
5. One concentric neutral strand is to continue up, under the current sensor where applicable, and connect to the drain wire probe on the elbow.

Figure 4.5 A three-phase pad-mounted distribution transformer.

Western Underground Committee Guide 2.13, entitled *Security for Padmounted Equipment Enclosures.*

The transformer shall have specific dimensions for Type-B transformers for loop-feed systems, as defined in the latest revision of ANSI C57.12.26. The transformer primary bushing wells and parking stands shall be located as shown in the standard.

The primary and secondary windings of the transformer shall be wound on a five-legged core and connected GrdY-GrdY.

The transformer shall be equipped with six primary bolted-on bushing wells rated 8.3 kV and 200 A, in accordance with the latest revision of ANSI/IEEE 386. The bushing wells shall be equipped with 8.3/14.4-kV 200-A load break bushing inserts to accommodate load break elbows for sectionalizing cables in a loop-feed primary system.

The transformer secondary bushings are to be equipped with copper studs, complete with bronze spade terminals.

The low-voltage neutral shall have a fully insulated bushing. A removable ground strap, sized for the rating of the transformer, shall be provided and connected between the neutral bushing terminal and a ground terminal in the compartment on the tank of the transformer.

The high- and low-voltage compartments shall each have a ground terminal on the transformer tank equipped with an eyebolt connector designed to accept a 1/0 bare copper wire. A ground pad shall be provided on the external surface of the transformer tank.

The high- and low-voltage terminals for the transformer windings shall be separated by a metal barrier.

The transformer shall be equipped with RTE Bay-O-Net, or approved equivalent, externally replaceable overload-sensing primary fuses, for coordinated overload and secondary fault protection, coordinated and in series with 8.3-kV partial-range current-limiting fuses. The current-limiting fuses shall not operate on secondary short circuits or transformer overloads. The current-limiting fuses shall be installed under oil inside the transformer tank. Each load-sensing and current-limiting fuse device shall be capable of withstanding a 15-kV system-recovery voltage across an open fuse.

The transformer shall be equipped with oil-drip shields directly below the Bay-O-Net fuse holders, to prevent oil dripping on the primary elbows or bushings during removal of the fuses.

The transformer tank shall have a removable cover or hand-hole of the tamper-resistant design.

A Qualitrol no. 202-032-01 automatic pressure relief device with an operation indication, or approved equivalent, shall be supplied on the transformer.

The transformer tank shall be equipped with a 1-inch oil drain valve with a built-in oil-sampling device.

The transformer shall be bar coded in accordance with the proposed EEI *Bar Coding Distribution Transformers: A Proposed EEI Guideline.* The manufacturer and serial number shall be bar coded on the nameplate. The serial number, manufacturer, and the company item ID (if given) shall be bar coded on a temporary label.

All transformers shall be filled with Type I insulating oil meeting the following requirements:

- All applicable requirements of ASTM D 3487, paragraph 82a.
- Minimum breakdown value of 28 kV per ASTM D 1816, paragraph 84a using a 0.4-inch gap.
- The PCB content shall be less than 2 parts per million, which shall be noted upon the nameplate of each unit. In addition, the manufacturer shall provide and install a "No PCB" identification on the tank wall, near the bottom of the tank centered below the low-voltage bushings. The decal shall be not less than 1 inch by 2 inch in size and shall indicate "No PCBs" in blue lettering on a white background, and be sufficiently durable to remain legible for the life of the transformer.

This specification can be changed to meet the kVA capacity and voltages required. Changes in the high voltage stated in this specification would require changes in the related items in the specification to obtain the proper insulation and surge-withstand capabilities.

Transformer Polarity

Subtractive polarity is when the current flows in the same direction in the two adjacent primary and secondary terminals (Fig. 4.6).
Additive polarity is when the current flows in the opposite direction in the two adjacent primary and secondary terminals (Fig. 4.7).

Polarity tests for overhead distribution transformers

Polarity may be further explained as follows: imagine a single-phase transformer having two high-voltage and three low-voltage external terminals (Fig. 4.8). Connect one high-voltage terminal to the adjacent low-voltage terminal, and apply a test voltage across the two high-voltage terminals. Then if the voltage across the unconnected high-voltage and low-voltage terminals is less than the voltage applied across the high-voltage terminals, the polarity is subtractive; if it is greater than the voltage applied across the high-voltage terminals, the polarity is additive.

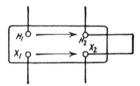

Figure 4.6 Sketch showing polarity markings and directions of voltages when polarity is subtractive.

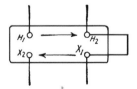

Figure 4.7 Sketch showing polarity markings and directions of voltages when polarity is additive.

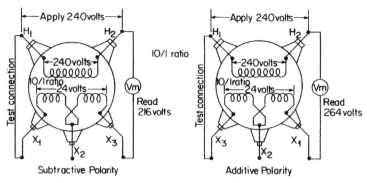

Figure 4.8 Polarity test diagrams for overhead distribution transformers.

Distribution transformer connections

Figures 4.9 through 4.17 show the following distribution transformer connections:

- Distribution transformer, pole-type single-phase two-wire service from a three-phase, three-wire delta ungrounded source (Fig. 4.9).

- Distribution transformer, pole-type single-phase three-wire service from a three-phase, three-wire, delta ungrounded source (Fig. 4.10).

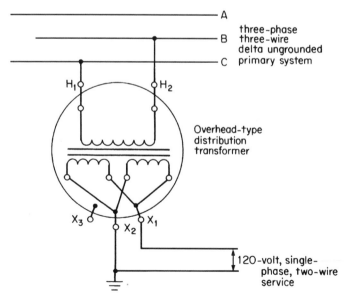

Figure 4.9 A single-phase overhead or pole-type distribution transformer connection for 120-volt two-wire secondary service. The transformer secondary coils are connected in parallel.

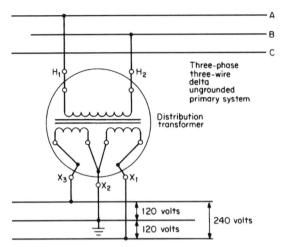

Figure 4.10   A single-phase distribution transformer connected to provide 120/240-volt three-wire single-phase service. The transformer secondary coils are connected in series.

- Distribution transformer, pole-type single-phase three-wire service from a three-phase, four-wire, wye-grounded source (Fig. 4.11).

- Secondary single-phase connections internal to the distribution transformer, pole type (Figs. 4.12 and 4.13).

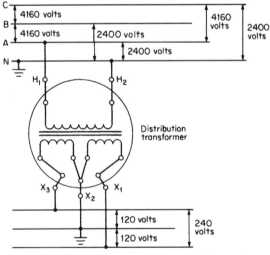

Figure 4.11   A single-phase distribution transformer is connected to provide 120/240-volt three-wire single-phase service. The primary winding is connected line to neutral or ground. Notice that the phase to phase voltage of 4160 volts is equal to the square root of 3 (1.73) times the phase to neutral voltage of 2400 volts (4160 = $\sqrt{3} \times 2400$).

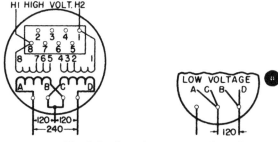

Figure 4.12 The left view shows connections for 120-/ 240-volt three-wire service. For 240-volt two-wire service, the neutral bushing conductor is not connected to the jumper between terminals B and C inside of the transformer tank. For 120-volt two-wire service, the connections are made, as shown in the right view. All connections are completed inside the transformer tank. (*Courtesy ABB Power T&D Company, Inc.*)

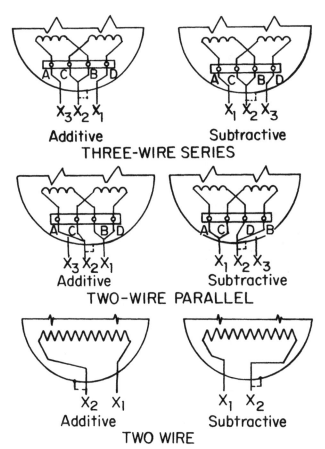

Figure 4.13 Standard low-voltage connections for a distribution transformer.

- Distribution transformers, pole-type three-phase grounded wye–grounded wye connection (Y-Y) (Fig. 4.14).

- Distribution transformers, pole-type three-phase delta-delta connection (Δ-Δ) (Fig. 4.15).

- Distribution transformers, pole-type three-phase delta-wye connection (Δ-Y) (Fig. 4.16).

- Distribution transformers, pole-type open delta-connected three phase (Fig. 4.17).

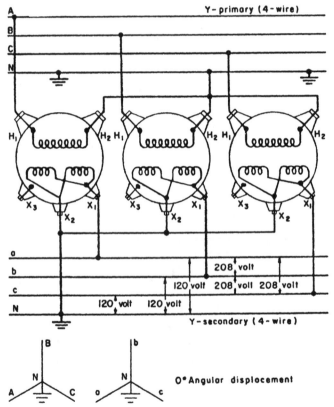

Figure 4.14   Three-phase Y-Y pole-type transformer connections used to obtain 208Y/120 volts three-phase from a three-phase, four-wire primary. If single-primary-bushing transformers are used, the H_2 terminal of the transformer primary winding would be connected to the tank of the transformer.

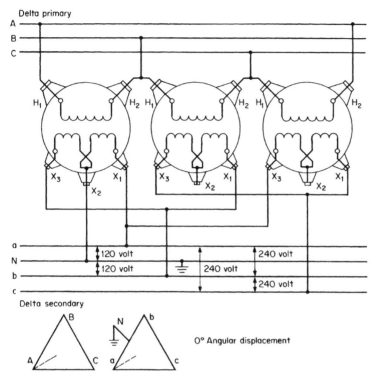

Delta primary

Delta secondary

0° Angular displacement

Figure 4.15 Three-phase delta-delta pole-type transformer connection used to obtain 240/120-volt single-phase service from a 240-volt three-phase service and a three-phase, three-wire primary. The transformer with the secondary connection to bushing X_2 provides a source for single-phase 120/240-volt service.

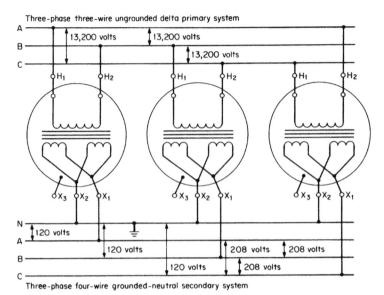

Three-phase three-wire ungrounded delta primary system

A

13,200 volts   13,200 volts

B

13,200 volts

C

N

120 volts

A

120 volts   208 volts   208 volts

B

120 volts   208 volts

C

Three-phase four-wire grounded-neutral secondary system

Figure 4.16 Three single-phase distribution transformers connected delta-wye (Δ-Y). Windings must be rated for the primary system and the desired secondary voltage.

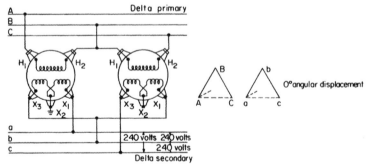

Delta primary

A
B
C

0° angular displacement

a
b
c

240 volts 240 volts
240 volts
Delta secondary

Figure 4.17 Two single-phase transformers connected into a bank to transform power from three-phase primary voltage to 240-volt three-phase. Primaries and secondaries are open-delta connected.

NOTES

NOTES

Wood-Pole Structures

Wood-Pole Types

- Southern yellow pine
- Western red cedar
- Douglas fir
- Western larch

Wood-Pole Preservatives

- Creosote
- Pentachlorophenal
- Chromated copper arsenates (CCA)

Wood-Pole Classifications, Dimensions, and Weights

Wood poles are commonly classified by length, top circumference, and circumference measured 6 ft from the butt end. The lengths vary in 5-ft steps, and the circumference at the top varies in 2-inches steps. There are lengths, in feet, of 20 to 120 ft and minimum top circumferences, in inches, of 15 to 95 inches.

The circumference measured 6 ft from the butt end determines to which class, numbered from 1 to 10, a pole of a given length and top circumference belongs. The classification from 1 to 10 determines the strength to resist loads applied 2 ft from the top of the pole. The H classes are applicable only for the brand western red cedar poles.

TABLE 5.1 Wood Pole, Minimum Pole Circumferences, Western Red Cedar, and Ponderosa Pine Poles

Dimensions of Western Red Cedar* and Ponderosa Pine Poles from ANSI 05.1-1992

Class	H-6	H-5	H-4	H-3	H-2	H-1	1	2	3	4	5	6	7	9	10
Minimum circumference at top (in)	39	37	35	33	31	29	27	25	23	21	19	17	15	15	12

Length of pole (ft)	Groundline† distance from butt (ft)	H-6	H-5	H-4	H-3	H-2	H-1	1	2	3	4	5	6	7	9	10
						Minimum circumference at 6 ft from butt (in)										
20	4.0	—	—	—	—	—	—	33.5	31.5	29.5	27.0	25.0	23.0	21.5	18.5	15.0
25	5.0	—	—	—	—	—	—	37.0	34.5	32.5	30.0	28.0	25.5	24.0	20.5	16.5
30	5.5	—	—	—	—	—	—	40.0	37.5	35.0	32.5	30.0	28.0	26.0	22.0	—
35	6.0	—	—	—	—	48.0	45.5	42.5	40.0	37.5	34.5	32.0	30.0	27.5	—	—
40	6.0	—	—	56.5	53.5	51.0	48.0	45.0	42.5	39.5	36.5	34.0	31.5	—	—	—
45	6.5	64.5	62.0	59.0	56.0	53.5	50.5	47.5	44.5	41.5	38.5	36.0	33.0	—	—	—
50	7.0	67.0	64.5	61.5	58.5	55.5	52.5	49.5	46.5	43.5	40.0	37.5	—	—	—	—
55	7.5	70.0	67.0	64.0	61.0	57.5	54.5	51.5	48.5	45.0	42.0	—	—	—	—	—
60	8.0	72.0	69.0	66.0	63.0	59.5	56.5	53.5	50.0	46.5	43.5	—	—	—	—	—
65	8.5	74.5	71.5	68.0	65.0	61.5	58.5	55.0	51.5	48.0	45.0	—	—	—	—	—
70	9.0	76.5	73.5	70.0	67.0	63.5	60.0	56.5	53.0	49.5	46.0	—	—	—	—	—
75	9.5	78.5	75.5	72.0	68.5	65.0	61.5	58.0	54.5	51.0	—	—	—	—	—	—
80	10.0	80.5	77.0	74.0	70.5	67.0	63.0	59.5	56.0	52.0	—	—	—	—	—	—
85	10.5	82.5	79.0	75.5	72.0	68.5	64.5	61.0	57.0	53.5	—	—	—	—	—	—
90	11.0	84.5	81.0	77.0	73.5	70.0	66.0	62.5	58.5	54.5	—	—	—	—	—	—
95	11.0	86.0	82.5	79.0	75.0	71.5	67.5	63.5	59.5	—	—	—	—	—	—	—
100	11.0	87.5	84.0	80.5	76.5	72.5	69.0	65.0	61.0	—	—	—	—	—	—	—
105	12.0	89.5	85.5	82.0	78.0	74.0	70.0	66.0	62.0	—	—	—	—	—	—	—
110	12.0	91.0	87.0	83.5	79.5	75.5	71.5	67.5	63.0	—	—	—	—	—	—	—
115	12.0	92.5	88.5	84.5	80.5	76.5	72.5	68.5	64.0	—	—	—	—	—	—	—
120	12.0	94.0	90.0	86.0	82.0	78.0	74.0	69.5	65.0	—	—	—	—	—	—	—
125	12.0	95.5	91.5	87.5	83.0	79.0	75.0	70.5	66.0	—	—	—	—	—	—	—

NOTE: Classes and lengths for which circumferences at 6 ft from the butt are listed in boldface type are the preferred standard sizes. Those shown in light type are for engineering purposes only.
*Dimensions of H classes are applicable for western red cedar only.
†The figures in this column are intended for use only when a definition of groundline is necessary in order to apply requirements relating to scars, straightness, etc.

TABLE 5.2 Wood Pole, Minimum Pole Circumferences, Douglas Fir, and Southern Pine Poles

Dimensions of Douglas fir (both types) and Southern Pine Poles from ANSI 05.1-1992

Length of pole (ft)	Groundline* distance from butt (ft)	Class 10	9	7	6	5	4	3	2	1	H-1	H-2	H-3	H-4	H-5	H-6
Minimum circumference at top (in)		12	15	15	17	19	21	23	25	27	29	31	33	35	37	39
		Minimum circumference at 6 ft from butt (in)														
20	4.0	14.0	17.5	19.5	21.0	23.0	25.0	27.0	29.0	31.0	—	—	—	—	—	—
25	5.0	15.0	19.5	21.5	23.0	25.5	27.5	29.5	31.5	33.5	—	—	—	—	—	—
30	5.5	—	20.5	23.5	25.0	27.5	29.5	32.0	34.0	36.5	—	—	—	—	—	—
35	6.0	—	—	25.0	27.0	29.0	31.5	34.0	36.5	39.0	41.5	43.5	—	—	—	—
40	6.0	—	—	—	28.5	31.0	33.5	36.0	38.5	41.0	43.5	46.0	48.5	51.0	—	—
45	6.5	—	—	—	30.0	32.5	35.0	37.5	40.5	43.0	45.5	48.5	51.0	53.5	56.0	58.5
50	7.0	—	—	—	—	34.0	36.5	39.0	42.0	45.0	47.5	50.5	53.0	55.5	58.5	61.0
55	7.5	—	—	—	—	—	38.0	40.5	43.5	46.5	49.5	52.0	55.0	58.0	60.5	63.5
60	8.0	—	—	—	—	—	39.0	42.0	45.0	48.0	51.0	54.0	57.0	59.5	62.5	65.5
65	8.5	—	—	—	—	—	40.5	43.5	46.5	49.5	52.5	55.5	58.5	61.5	64.5	67.5
70	9.0	—	—	—	—	—	41.5	45.0	48.0	51.0	54.0	57.0	60.5	63.5	66.5	69.0
75	9.5	—	—	—	—	—	—	46.0	49.0	52.5	55.5	59.0	62.0	65.0	68.0	71.0
80	10.0	—	—	—	—	—	—	47.0	50.5	54.0	57.0	60.0	63.5	66.5	69.5	72.5
85	10.5	—	—	—	—	—	—	48.0	51.5	55.0	58.5	61.5	65.0	68.0	71.5	74.5
90	11.0	—	—	—	—	—	—	49.0	53.0	56.0	59.5	63.0	66.5	69.5	73.0	76.0
95	11.0	—	—	—	—	—	—	—	54.0	57.0	61.0	64.5	67.5	71.0	74.5	77.5
100	11.0	—	—	—	—	—	—	—	55.0	58.5	62.0	65.5	69.0	72.5	76.0	79.0
105	12.0	—	—	—	—	—	—	—	56.0	59.5	63.0	67.0	70.5	74.0	77.0	80.5
110	12.0	—	—	—	—	—	—	—	57.0	60.5	64.5	68.0	71.5	75.0	78.5	82.0
115	12.0	—	—	—	—	—	—	—	58.0	61.5	65.5	69.0	72.5	76.5	80.0	83.5
120	12.0	—	—	—	—	—	—	—	59.0	62.5	66.5	70.0	74.0	77.5	81.0	85.0
125	12.0	—	—	—	—	—	—	—	59.5	63.5	67.5	71.0	75.0	78.5	82.5	86.0

NOTE: Classes and lengths for which circumferences at 6 ft from the butt are listed in boldface type are the preferred standard sizes. Those shown in light type are included for engineering purposes only.

*The figures in this column are intended for use only when a definition of groundline is necessary in order to apply requirements relating to scars, straightness, etc.

TABLE 5.3 Weights of Wood Poles

Size/Class	Weight	Material	Size/Class	Weight	Material	Size/Class	Weight	Material
30-6	556	Southern Yellow Pine	55-1	3258	Southern Yellow Pine	70-5	4278	Douglas Fir
	415	Red Pine		1937	Red Pine	70-H1	4784	Douglas Fir
	328	Western Red Cedar		1674	Western Red Cedar	70-H2	5888	Douglas Fir
35-4	841	Southern Yellow Pine	55-2	2826	Southern Yellow Pine	70-H4	6532	Douglas Fir
	748	Red Pine		1650	Red Pine	70-H5	7268	Douglas Fir
	549	Western Red Cedar		1464	Western Red Cedar	70-H6	3981	Douglas Fir
35-5	814	Southern Yellow Pine	55-3	2066	Southern Yellow Pine	75-1	3511	Douglas Fir
	641	Red Pine		1443	Red Pine	75-2	4278	Douglas Fir
	482	Western Red Cedar		1277	Western Red Cedar	75-3	4784	Douglas Fir
40-2	1430	Southern Yellow Pine	60-1	2782	Southern Yellow Pine	75-H1	5244	Douglas Fir
	1217	Red Pine		1912	Red Pine	75-H2	5888	Douglas Fir
	899	Western Red Cedar		2477	Western Red Cedar	75-H3	6532	Douglas Fir
40-4	1166	Southern Yellow Pine	60-2	1669	Southern Yellow Pine	75-H4	7268	Douglas Fir
	791	Red Pine		2153	Red Pine	75-H5	4503	Douglas Fir
	675	Western Red Cedar		1453	Western Red Cedar	75-H6	4042	Douglas Fir
45-1	1977	Southern Yellow Pine	60-3	3404	Southern Yellow Pine	80-1	5244	Douglas Fir
	1661	Red Pine		2156	Red Pine	80-2	5842	Douglas Fir
	1224	Western Red Cedar		3772	Western Red Cedar	80-H1	6486	Douglas Fir
45-2	1876	Southern Yellow Pine	60-H1	2400	Douglas Fir	80-H2	7176	Douglas Fir
	1432	Red Pine	60-H2	4186	Douglas Fir	80-H3	7968	Douglas Fir
	1070	Western Red Cedar	60-H3	2643	Douglas Fir	80-H4	8832	Douglas Fir
45-3	1623	Southern Yellow Pine	60-H6	5750	Douglas Fir	80-H5	4860	Douglas Fir
	1242	Red Pine	65-1	3374	Southern Yellow Pine	80-H6	4385	Douglas Fir
	935	Western Red Cedar		3150	Red Pine	85-1	5750	Douglas Fir
50-1	2348	Southern Yellow Pine		2806	Western Red Cedar	85-2	6394	Douglas Fir
	1937	Red Pine	65-H1	3818	Douglas Fir	85-H1	7084	Douglas Fir
	1448	Western Red Cedar	65-H2	4232	Douglas Fir	85-H2	7866	Douglas Fir
50-2	2222	Southern Yellow Pine	65-H3	4738	Douglas Fir	85-H3	9706	Douglas Fir
	1650	Red Pine	65-H4	4846	Douglas Fir	85-H4	5250	Douglas Fir
	1271	Western Red Cedar	65-H5	5842	Douglas Fir	85-H5	6302	Douglas Fir
50-3	1925	Southern Yellow Pine	65-H6	6486	Douglas Fir	85-H6		Douglas Fir
	1443	Red Pine	70-1	3553	Red Pine	90-1		Douglas Fir
	1098	Western Red Cedar	70-2	3144	Western Red Cedar	90-H1		Douglas Fir

Pole Hole Depths

TABLE 5.4   Recommended Pole-Setting Depths in Soil
and Rock for Various Lengths of Wood Poles

Length of pole, ft	Setting depth in soil, ft	Setting depth in rock, ft
25	5.0	3.5
30	5.5	3.5
35	6.0	4.0
40	6.0	4.0
45	6.5	4.5
50	7.0	4.5
55	7.5	5.0
60	8.0	5.0
65	8.5	6.0
70	9.0	6.0
75	9.5	6.0
80	10.0	6.5
85	10.5	7.0
90	11.0	7.5
95	11.0	7.5
100	11.0	7.5
105	12.0	8.0
110	12.0	8.0
115	12.0	8.0
120	13.0	8.5
125	13.0	8.5
130	13.0	8.5
135	14.0	9.0

SOURCE: Commonwealth Edison Company.

TABLE 5.5   Recommended Pole-Setting Depths in Soil and Rock

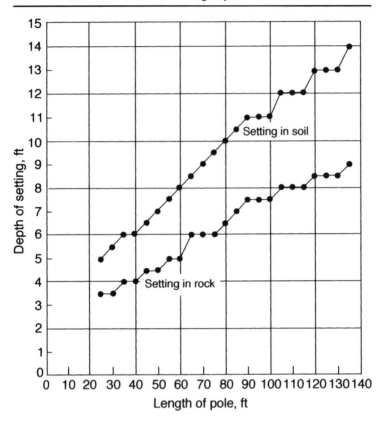

Wood-Pole Construction Standards Examples

The following examples demonstrate the varied applications utilizing the wood-pole in the construction of distribution, sub-transmission, and transmission structures (Figs. 5.1 through 5.9).

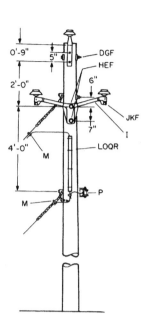
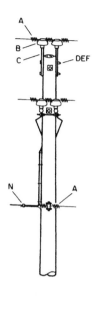

Item	No. required
A. Preformed tie wire size required	7
B. 1" Pin insulator	6
C. 1" Pole-top pin	2
D. 5/8" × 10" Bolt	3
E. 5/8" Round washer	2
F. 5/8" Palnut	9
G. 2 1/2" Curved washer	1
H. 5/8" × 12" Bolt	2
I. Two-phase brackets 48" spacing	2
J. 5/8" × 14" Double arming bolt	2
K. 2 1/4" Square washer	8
L. #6 Copper wire	1#
M. Split-bolt connector, size required	3
N. Compression connector, size required	1
O. Molding, ground wire	5'
P. Single-clevis insulator	1
Q. Staples, molding	6
R. Staples, fence	6

Figure 5.1 Three-phase 13.2Y/7.6-kV armless-type
wood-pole structure (angle pole 6° to 20°).

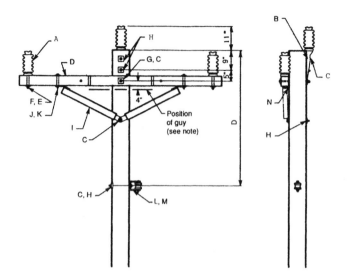

Conductor clearances		
Span length (ft)	Pole length (ft)	Neutral location D
200 & less	All	4'0"
201–270	35	6'0"
271–310	40 & greater	7'6"

NOTE: Guy for angles greater than 1°.

Item	Description
A	Insulator, 35 kV, post-type, clamp top.
B	Bracket for post-type insulator.
C	Bolt, 3/4" diameter, length required.
D	Crossarm, wood, 10' long, 4 pin.
E	Washer, round for 5/8" diameter bolt.
F	Bolt, stud, 5/8" diameter, 7" long for line post insulator.
G	Washer, square, flat for 3/4" diameter bolt.
H	Washer, square, curved for 3/4" diameter bolt.
I	Brace, crossarm, wood.
J	Bolt, carriage, 1/2" diameter, 6" long.
K	Washer, round for 1/2" diameter bolt.
L	Insulator, spool, 3" diameter.
M	Clevis for spool insulator.
N	Gain 3" X 4".

Figure 5.2 Drawing of a 34.5Y/19.9-kV three-phase tangent distribution-circuit structure for installation in a straight portion of the line or where any angle in the line is small.

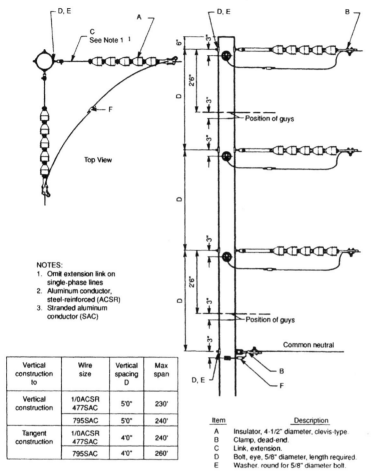

NOTES:
1. Omit extension link on single-phase lines
2. Aluminum conductor, steel-reinforced (ACSR)
3. Stranded aluminum conductor (SAC)

Vertical construction to	Wire size	Vertical spacing D	Max span
Vertical construction	1/0ACSR 477SAC	5'0"	230'
	795SAC	5'0"	240'
Tangent construction	1/0ACSR 477SAC	4'0"	240'
	795SAC	4'0"	260'

Item	Description
A	Insulator, 4-1/2" diameter, clevis-type.
B	Clamp, dead-end.
C	Link, extension.
D	Bolt, eye, 5/8" diameter, length required.
E	Washer, round for 5/8" diameter bolt.

Figure 5.3   Drawing of a 34.5Y/19.9-kV three-phase angle distribution-circuit structure.

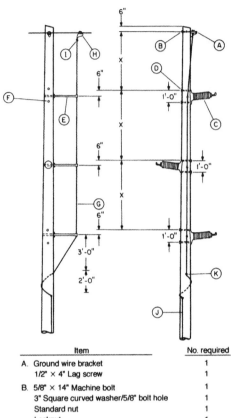

Item	No. required
A. Ground wire bracket	1
1/2" × 4" Lag screw	1
B. 5/8" × 14" Machine bolt	1
3" Square curved washer/5/8" bolt hole	1
Standard nut	1
Locknut	1
C. 88 kV Horizontal post insulator, gray color	3
D. 3/4" × 12" Machine bolt	2
3/4" × 14" Machine bolt	3
3/4" × 16" Machine bolt	1
3" Square curved washer/3/4" bolt hole	6
Standard nut	6
MF Locknut	6
E. Fiber glass downlead bracket	3
F. 5/8" × 12" Machine bolt	3
Standard nut	3
2 1/4" Square washer/5/8" bolt hole	6
G. 6A Copperweld groundwire	
H. Two-bolt connector	1
I. Split-bolt connector	1
J. Ground-wire molding 1/2" groove	
K. Bronze staple, for 1/2" plastic molding	

NOTE: Phase spacing (dimension X) varies with span lengths, conductor size, conductor sag, and midspan clearances.

Figure 5.4 A 69-kV single-circuit armless-type tangent structure.

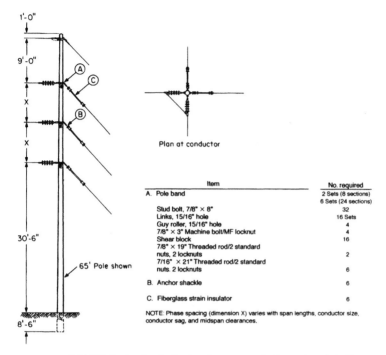

Figure 5.5 A 69-kV single-circuit dead-end structure (45° to 90° angle in line).

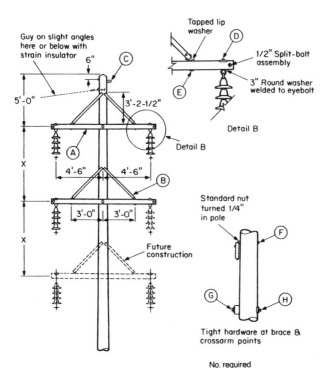

Item	No. required	
	Single circuit	Double circuit
A. Crossarm	2	3
B. Crossarm brace (1 left, 1 right)	2 Pair	3 Pair
C. Shield-wire support	1	1
Washer nut and standard nut	1	1
D. 5/8" × 6 1/2" Eyebolt	3	6
Saddle washer assembly		
Standard nut		
E. 5/8" × 7" Washer head bolt	3	6
Standard nut	3	6
MF locknut	3	6
F. 5/8" × 12" Washer head bolt	1	1
5/8" × 14" Washer head bolt	1	2
Standard nut	2	3
G. 3/4" × 16" Washer head bolt	1	1
3/4" × 18" Washer head bolt	1	2
H. 3/4" Nut and locknut	2	3
Curved guide washer staked	2	3

NOTE: Phase spacing (dimension X) varies with span lengths, conductor size, conductor sag, and midspan clearances. Spacing will be determined in accordance with appropriate design standard.

Figure 5.6 A 69-kV double-circuit crossarm construction tangent structure (0° to 30° angle).

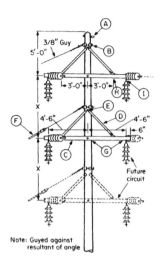

	No. required	
	Single	Double
Item	circuit	circuit
A. Shield-wire support	1	1
Washer nut and standard nut	1	1
B. Pole band	2 Sets	3 Sets
	(8 sections)	(12 sections)
Stud bolt, 3/4" × 6"	8	12
Links	2 Sets	3 Sets
Guy roller, 15/16" hole	2	3
3/4" × 2 1/2" Machine bolt/MF	2	3
locknut		
C. Double arm crossarm assembly	2	3
D. Crossarm brace (1 left, 1 right)	4 Pair	6 Pair
E. 5/8" × 14" Double arming bolt	2	2
5/8" × 16" Double arming bolt	0	1
4 Standard nuts per bolt		
F. Guy strain insulator, fiber glass	As required	
G. 3/4" × 28" Double arming bolts	2	3
2 Washer nuts and 2 standard	4	6
nuts per bolt		
3/4" × 28" Double arming bolts	8	12
2 Washer nuts and 6 standard	16	24
nuts per bolt		
Dead-end Tee	8	12
H. 5/8" × 6" Washer head bolt	8	12
with lip washer		
I. 5/8" × 6" Eyebolt with flange	3	6
washer and standard nut		

NOTE: Phase spacing (dimension X) varies with span
lengths, conductor size, conductor sag, and midspan
clearances. Spacing will be determined in accordance
with appropriate design standard.

Figure 5.7   A 69-kV double-circuit crossarm construction for a 7° to 30° angle
structure.

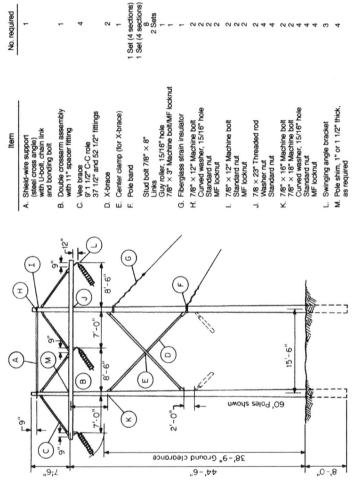

Item	No. required
A. Shield-wire support (steel cross angle) with U-bolt, chain link and bonding bolt	1
B. Double crossarm assembly with 11" spacer fitting	1
C. Vee brace 9' 1 1/2" C-C role 37 1/2" and 52 1/2" fittings	4
D. X-brace	2
E. Center clamp (for X-brace)	1
F. Pole band	1 Set (4 sections)
	1 Set (4 sections)
Stud bolt 7/8" × 8"	8
Links	2 Sets
Guy roller, 15/16" hole	1
7/8" × 3" Machine bolt/MF locknut	1
G. Fiberglass strain insulator	1
H. 7/8" × 12" Machine bolt	2
Curved washer, 15/16" hole	2
Standard nut	2
MF locknut	2
I. 7/8" × 12" Machine bolt	2
Standard nut	2
MF locknut	2
J. 7/8 × 23" Threaded rod	2
Washer nut	4
Standard nut	4
K. 7/8" × 16" Machine bolt	2
7/8" × 18" Machine bolt	2
Curved washer, 15/16" hole	4
Standard nut	4
MF locknut	4
L. Swinging angle bracket	3
M. Pole shim, 1" or 1 1/2" thick, as required	4

Figure 5.8 A 161-kV H-frame structure for installation when the line has an angle of $1\frac{1}{2}°$ to $6°$.

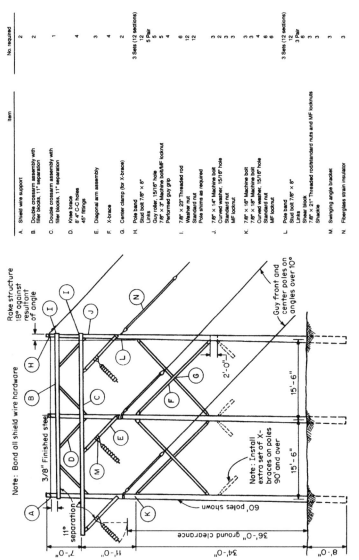

Item	No. required
A. Shield wire support	2
B. Double crossarm assembly with filler blocks, 11" separation	2
C. Double crossarm assembly with filler blocks, 11" separation	1
D. Knee brace 8' 4" C-C holes 45° fittings	4
E. Diagonal arm assembly	3
F. X-brace	4
G. Center clamp (for X-brace)	2
H. Pole band Stud bolt 7/8" × 8"	3 Sets (12 sections)
Links	12
Guy roller, 15/16" hole	5 Pair
7/8" × 3" Machine bolt/MF locknut	5
Performed guy grip	4
I. 7/8" × 23" Threaded rod	6
Washer nut	12
Standard nut	12
Pole shims as required	
J. 7/8" × 14" Machine bolt	3
Curved washer, 15/16" hole	2
Standard nut	2
MF locknut	3
K. 7/8" × 16" Machine bolt	3
7/8" × 18" Machine bolt	4
Curved washer, 15/16" hole	6
Standard nut	6
MF locknut	6
L. Pole band Stud bolt 7/8" × 8"	3 Sets (12 sections)
Links	12
Shear block	3 Pair
7/8" × 21" Threaded rod/standard nuts and MF locknuts	6
Shackle	3
M. Swinging angle bracket	3
N. Fiberglass strain insulator	3

Note: Bond all shield wire hardware

Rake structure 18° against resultant of angle

3/8" Finished steel

11° separation

60' poles shown

36'-0" ground clearance

Guy front and center poles on angles over 10°

Note: Install extra set of X-braces on poles 90' and over

Figure 5.9 A 161-kV line three-pole 6° to 17½° angle structure.

NOTES

6

Guying

Figure 6.1 shows an anchor and down guy assembly.

Definitions

- **Guy**   A brace or cable fastened to the pole to strengthen it and keep it in normal position.

- **Anchor or down guy**   A wire running from the attachment near the top of the pole to a rod and anchor installed in the ground.

- **Terminal guy**   An anchor or down guy used at the ends of pole lines in order to counter balance the pull of the line conductors.

- **Line guys or storm guys**   Guys installed at regular intervals to protect transmission facilities and limit the damage if a conductor breaks.

- **Span guy**   A guy wire installed from the top of a pole to the top of an adjacent pole to remove the strain from the line conductors.

- **Head guy**   A guy wire running from the top of a pole to a point below the top of the adjacent pole.

- **Arm guy**   A guy wire running from one side of a crossarm to the next pole.

- **Stub guy**   A guy wire installed between a line pole and a stub pole on which there is no energized equipment.

- **Push guy**   A pole used as a brace to a line pole.

- **Guy wire**   Galvanized steel wire or Alumoweld wire used in a down guy.

- **Anchor**   The device used to counteract the unbalanced forces on electric transmission and distribution line structures.

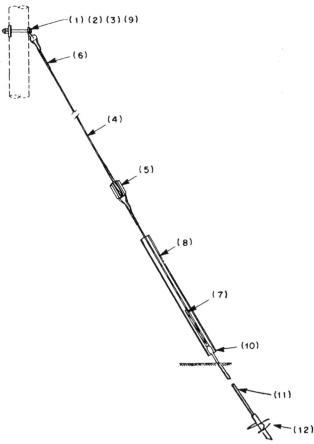

Figure 6.1 Anchor and down guy assembly. (1) Galvanized machine bolt with nut; (2) locknut; (3) square curved galvanized washer; (4) galvanized steel guy wire; (5) porcelain guy strain insulator; (6) prefabricated guy dead-end grip; (7) prefabricated guy dead-end grip; (8) plastic guy guard; (9) angle thimbleye; (10) eyenut; (11) steel anchor rod; and (12) power-installed screw anchor.

Tables 6.1 through 6.3 show appropriate data for guying materials.

Conductor Breaking Strengths

Tables 6.4 through 6.10 provide information relating to the breaking strength of conductors.

TABLE 6.1 Guy Strand Sizes and Strengths

Number of wires in strand	Diameter, in.	Minimum breaking strength of strand, lb.					
		Utilities grade	Common grade	Siemens-Martin grade	High-strength grade	Extra-high-strength grade
7	9/32	4,600	2,570	4,250	6,400	8,950
7	5/16	6,000	3,200	5,350	8,000	11,200
7	3/8	11,500	4,250	6,950	10,800	15,400
7	7/16	18,500	5,700	9,350	14,500	20,800
7	1/2	25,000	7,400	12,100	18,800	26,900
7	9/16		9,600	15,700	24,500	35,000
7	5/8		11,600	19,100	29,600	42,400
19	1/2			12,700	19,100	26,700
19	9/16			16,100	24,100	33,700
19	5/8			18,100	28,100	40,200
19	3/4			26,200	40,800	58,300
19	7/8			35,900	55,800	79,700
19	1			47,000	73,200	104,500
37	1			46,200	71,900	102,700
37	1 1/8			58,900	91,600	130,800
37	1 1/4			73,000	113,600	162,200

TABLE 6.2 Holding Power of Commonly Used
Manufactured-Type Anchors

Anchor		Holding strength, lb.	
Type	Size, in.	Poor soil	Average soil
Expansion	8	10,000	17,000
Expansion	10	12,000	21,000
Expansion	12	16,000	26,500
Screw or helix*	8	6,000	15,000
Screw or helix	11⁵/₁₆	9,500	15,000
Screw or helix	32	10,000	23,000
Screw or helix	34	15,500	27,000
Never-creep		21,000	34,000
Cross-plate		18,000	30,000

*Screw or helix anchors power-installed.

TABLE 6.3 Anchor Rod Strength

Anchor rod nominal diameter, in.	Ultimate tensile strength, lb/sq. in.	Yield point, lb/sq. in.	Full rod section		Threaded section	
			Ultimate load, lb.	Yield load, lb.	Ultimate load, lb.	Yield load, lb.
1/2	70,000	46,500	13,700	9,120	12,750	8,450
5/8	70,000	46,500	20,400	13,500	18,700	12,400
3/4	70,000	46,500	29,600	19,600	27,100	18,000
1	70,000	46,500	53,200	35,300	42,400	28,100
1 high strength	90,000	59,500	68,200	45,100	54,500	36,000
1¹/₄	70,000	46,500	85,800	57,000	67,800	45,000
1¹/₄ high strength	90,000	59,500	110,000	73,000	87,200	57,600

TABLE 6.4 Hard-Drawn Copper Wire, Solid

Size, AWG	Diameter, in.	Breaking strength, lb.
10	0.102	529
8	0.129	826
6	0.162	1,280
4	0.204	1,970
2	0.258	3,003
1	0.289	3,688
1/0	0.325	4,517
2/0	0.365	5,519
3/0	0.410	6,722
4/0	0.460	8,143

TABLE 6.5   Stranded Hard-Drawn Copper Wire

Size, AWG or cir. mils	No. of strands	Conductor diameter, in.	Breaking strength, lb.
6	7	0.184	1,288
4	7	0.232	1,938
2	7	0.292	3,045
1	7	0.328	3,804
1/0	7	0.368	4,752
2/0	7	0.414	5,926
3/0	7	0.464	7,366
4/0	7	0.522	9,154
4/0	19	0.528	9,617
250	19	0.574	11,360
300	19	0.628	13,510
350	19	0.679	15,590
400	19	0.726	17,810
450	19	0.770	19,750
500	19	0.811	21,950
500	37	0.813	22,500
600	37	0.891	27,020
700	61	0.964	31,820
750	61	0.998	34,090
800	61	1.031	36,360
900	61	1.094	40,520
1,000	61	1.152	45,030
1,250	91	1.289	56,280
1,500	91	1.412	67,540
1,750	127	1.526	78,800
2,000	127	1.632	90,050
2,500	127	1.824	111,300
3,000	169	1.998	134,400

TABLE 6.6   Copperweld Copper, Three Wire

	Type A	
Conductor	Cable diameter, in.	Breaking load, lb.
2A	0.366	5876
3A	0.326	4810
4A	0.290	3938
5A	0.258	3193
5D	0.310	6035
6A	0.230	2585
6D	0.276	4942
7A	0.223	2754
7D	0.246	4022
8A	0.199	2233
8D	0.219	3256
9½ D	0.174	1743

TABLE 6.7   Copperweld Standard Wire

| Size | Diameter, in. | Breaking strength, lb. | |
		High strength	Extra high strength
13/16 (19 No. 6)	0.810	45,830	55,530
21/32 (19 No. 8)	0.642	31,040	37,690
5/8 (7 No. 4)	0.613	24,780	29,430
1/2 (7 No. 6)	0.486	16,890	20,460
3/8 (7 No. 8)	0.385	11,440	13,890
5/16 (7 No. 10)	0.306	7,758	9,196
3 No. 6	0.349	7,639	9,754
3 No. 8	0.277	5,174	6,282
3 No. 10	0.220	3,509	4,160
3 No. 12	0.174		

TABLE 6.8   Aluminum Stranded Wire

Size, cir. mils or AWG	No. of strands	Diameter, in.	Ultimate strength, lb.
6	7	0.184	528
4	7	0.232	826
3	7	0.260	1,022
2	7	0.292	1,266
1	7	0.328	1,537
1/0	7	0.368	1,865
2/0	7	0.414	2,350
3/0	7	0.464	2,845
4/0	7	0.522	3,590
266,800	7	0.586	4,525
266,800	19	0.593	4,800
336,400	19	0.666	5,940
397,500	19	0.724	6,880
477,000	19	0.793	8,090
477,000	37	0.795	8,600
556,500	19	0.856	9,440
556,500	37	0.858	9,830
636,000	37	0.918	11,240
715,500	37	0.974	12,640
715,500	61	0.975	13,150
795,000	37	1.026	13,770
795,000	61	1.028	14,330
874,500	37	1.077	14,830
874,500	61	1.078	15,760
954,000	37	1.124	16,180
954,000	61	1.126	16,860
1,033,500	37	1.170	17,530
1,033,500	61	1.172	18,260
1,113,000	61	1.216	19,660
1,272,000	61	1.300	22,000
1,431,000	61	1.379	24,300
1,590,000	61	1.424	27,000
1,590,000	91	1.454	28,100

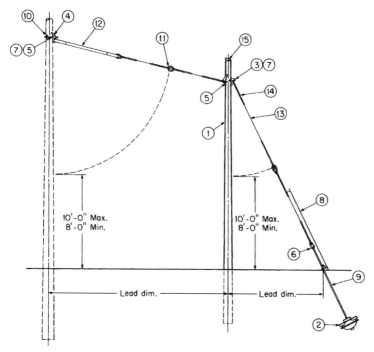

Figure 6.2 Stub guy and down guy assembly. (1) Pole stub; (2) anchor, expansion; (3) guy attachment, thimbleeye; (4) guy attachment, eye type; (5) bolt, standard machine, with standard nut; (6) clamp, guy bond, for 3/4-in. rod, twin eye; (7) locknut, 3/4-in. galvanized; (8) protector, guy; (9) rod, twin eye, 3/4-in. × 9 ft., with two standard nuts; (10) washer, 4-in. × 4-in. square curved; (11) insulator, strain, porcelain; (12) insulator, strain, fiberglass guy; (13) wire, guy; (14) grip, guy wire; and (15) pole topper.

Figure 6.2 shows a stub guy and down guy assembly.

Guy Size Determination

- Weight of the conductor
- The size and weight of crossarms and insulators
- Wind pressures on poles and conductors
- Strains caused by the contour of the earth
- Line curvatures
- Pole heights
- Dead-end loads
- Vertical load because of sleet and ice

TABLE 6.9   ACSR

Size, cir. mils or AWG	No. of strands	Diameter, in.	Ultimate strength, lb.
6	6 × 1	0.198	1,170
6	6 × 1	0.223	1,490
4	6 × 1	0.250	1,830
4	7 × 1	0.257	2,288
3	6 × 1	0.281	2,250
2	6 × 1	0.316	2,790
2	7 × 1	0.325	3,525
1	6 × 1	0.355	3,480
1/0	6 × 1	0.398	4,280
2/0	6 × 1	0.447	5,345
3/0	6 × 1	0.502	6,675
4/0	6 × 1	0.563	8,420
266,800	18 × 1	0.609	7,100
266,800	6 × 7	0.633	9,645
266,800	26 × 7	0.642	11,250
300,000	26 × 7	0.680	12,650
336,400	18 × 1	0.684	8,950
336,400	26 × 7	0.721	14,050
336,400	30 × 7	0.741	17,040
397,500	18 × 1	0.743	10,400
397,500	26 × 7	0.783	16,190
397,500	30 × 7	0.806	19,980
477,000	18 × 1	0.814	12,300
477,000	24 × 7	0.846	17,200
477,000	26 × 7	0.858	19,430
477,000	30 × 7	0.883	23,300
556,500	26 × 7	0.914	19,850
556,500	30 × 7	0.927	22,400
556,500	30 × 7	0.953	27,200
605,000	24 × 7	0.953	21,500
605,000	26 × 7	0.966	24,100
605,000	30 × 19	0.994	30,000
636,000	24 × 7	0.977	22,600
636,000	26 × 7	0.990	25,000
636,000	30 × 19	1.019	31,500
666,600	24 × 7	1.000	23,700
715,500	54 × 7	1.036	26,300
715,500	26 × 7	1.051	28,100
715,500	30 × 19	1.081	34,600
795,000	54 × 7	1.093	28,500
795,000	26 × 7	1.108	31,200
795,000	30 × 19	1.140	38,400
874,500	54 × 7	1.146	31,400
900,000	54 × 7	1.162	32,300
954,000	54 × 7	1.196	34,200
1,033,500	54 × 7	1.246	37,100
1,113,000	54 × 19	1.292	40,200
1,272,000	54 × 19	1.382	44,800
1,431,000	54 × 19	1.465	50,400
1,590,000	54 × 19	1.545	56,000

Computing the load on the guy

When the line is dead-ended requiring a terminal guy:

S = ultimate strength of the conductor
N = number of conductors

$$S \times N = line\ load$$

Example: If three 1/0 ACSR conductors are dead-ended on a pole, calculate the line load.

Solution: The ultimate strength for one 1/0 ACSR conductor is 4280 lbs., as found in Table 6.9:

$$S \times N = line\ load$$
$$4280 \times 3 = 12,840\ \text{lbs.}$$

When the line makes an angle requiring a side guy:

S = ultimate strength of the conductor
N = number of conductors
M = multiplication factor for the angle change of line direction (Table 6.10)

$$S \times N \times M = line\ load$$

Example: If three 1/0 ACSR conductors have a side angle of 15°, calculate the line load.

Solution: The ultimate strength for one 1/0 ACSR conductor is 4280 lbs. as found in Table 6.9. The multiplication factor for a side angle of 15° is 0.262, as found in Table 6.10:

$$S \times N \times M = line\ load$$
$$4280 \times 3 \times 0.262 = 3364\ \text{lbs.}$$

TABLE 6.10 Line-Angle Change Multiplication Factor

Angle change of line direction, deg.	Multiplication factor, M
15	0.262
30	0.518
45	0.766
60	1.000
75	1.218
90	1.414

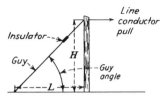

Figure 6.3 A sketch of the guy installation, where H is the height of guy attachment to the pole and L is the lead or distance from the pole to the anchor rod.

Computing the guy angle factor

The guy angle (Fig. 6.3) is calculated by applying the trigonometric laws of a right triangle. The law states that the tangent of the angle Θ or the guy angle factor equals the height divided by the length. The height in the case of a guyed pole is the height of the guy attachment on the pole. The length in the case of a guyed pole is the distance of the anchor from the base of the pole:

$$Guy\ angle\ factor = Tan\ \Theta = Height/Length$$

The angle Θ equals the inverse tangent of the height divided by the length:

$$\Theta = Tan^{-1}\ (Height/Length)$$

Example: If the attachment point on the pole is 30' from the ground and the anchor lead length is 15', calculate the guy angle factor.

Solution: The height in the case of a guyed pole is the height of the guy attachment on the pole. The length in the case of a guyed pole is the distance of the anchor from the base of the pole. Thus, for the example, the guy angle factor is shown to be

$$Guy\ angle\ factor = Tan\ \Theta = Height/Length = 30/15 = 2.0$$

TABLE 6.11   Guy Angle Factor

Guy angle factor Tan Θ	Height	Length	Θ
3.0	30	10	71.5°
2.0	30	15	63.5°
1.732	26	15	60°
1.5	30	20	56.3°

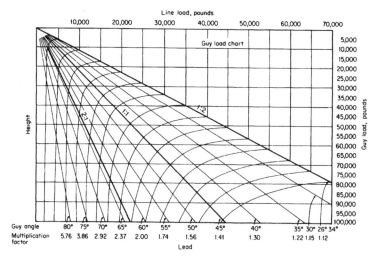

Figure 6.4   Chart for determining guy load if line load and guy angle are known. (*Courtesy A. B. Chance Co.*)

The corresponding guy angle is 63.5°, as found in Table 6.11.

Example: If the line load is calculated to be 10,000 lbs. and the guy angle is 60°, determine from the chart, the guy load on the anchor.

Solution: The line load is listed horizontally across the top of the chart. Reading the value of 10,000 lbs. and going down along this line until it intersects (or crosses) the guy angle diagonal line for 60°, and then reading across to the value of 20,000 lbs. on the right side of the chart results in the guy load on the anchor. See Fig. 6.4 for a complete guy load table.

NOTES

Lightning and Surge Protection

Figure 7.1 shows the number of electrical storm days per year.

Definitions

- **Lightning** A current of electricity flowing from one cloud to another or between the cloud and the earth.

- **Surge** Large overvoltages that develop suddenly on electric transmission and distribution circuits.

- **Shield wire** A ground wire mounted above the other line conductors, providing protection from direct lightning strokes to the other line conductors of the circuit.

- **Lightning or surge arresters** Devices that prevent high voltages, which would damage equipment, from building up on a circuit, by providing a low-impedance path to ground for the current from lightning or transient voltages, and then restoring normal circuit conditions.

- **Basic insulation level (BIL)** The crest value of voltage of a standard full-impulse voltage wave that the insulation will withstand without failure.

- **Ground lead isolating device** A heat-sensitive device that separates the ground lead from a defective arrester. When the ground lead is separated from the arrester, the lightning arrester will be isolated from ground, permitting normal circuit operation.

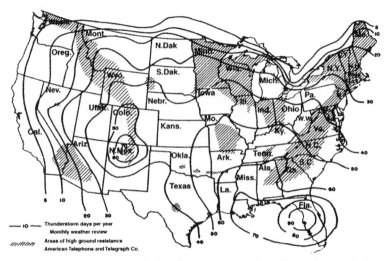

Figure 7.1 Lines are marked with the average number of lightning-producing storm days occurring in each area per year. (*Courtesy American Telephone and Telegraph Co.*)

Classes of Surge Arresters

- **Secondary**   Voltage ratings of 175 and 650 volts.

- **Distribution**   Voltage ratings of 3, 6, 9, 10, 12, 15, 18, 21, 25, 27, and 30 kV.

- **Intermediate**   Voltage ratings of 3 kV through 120 kV.

- **Station**   Voltage ratings of 3 kV and above.

Valve Block Types of Surge Arresters

Silicon-carbide arresters are surge arresters with silicon-carbide blocks that have gaps in a sealed porcelain casing and grading resistors or capacitors to control the sparkover voltages of the gaps (Tables 7.1 through 7.4). The silicon-carbide valve blocks function as a nonlinear resistor that passes large surge currents and limits the power-follow current.

TABLE 7.1   Basic Insulation Levels for Distribution Systems

Voltage	Basic insulation level (BIL)
5 kV	70
15 kV	95
25 kV	125
35 kV	150

TABLE 7.2 Typical Application Data for Silicon-Carbide Distribution-Class, Nonlinear Resistor Valve-Type Lightning Arresters with Gaps (Electrical Characteristics)

Arrester rating maximum line to ground, kV rms	Minimum 60-Hz sparkover, kV rms	Maximum 1½ × 40 wave kV crest	Maximum impulse sparkover ANSI front-of-wave kV crest, maximum	Discharge voltage in kV crest for discharge currents of 8 × 20 microsecond waveshape with following maximum crest amplitudes				
				1500-A kV crest, maximum	5000-A kV crest, maximum	10,000-A kV crest, maximum	20,000-A kV crest, maximum	65,000-A kV crest, maximum
3	6	14	17	10	12.4	13.8	15.5	19
6	11	26	33	19	23	26	29	35
9	18	38	47	30	36.5	41	46	56
10	18	38	48	30	36.5	41	46	56
12	23.5	49	60	38	46	52	58	71
15	27	50	75	45	55	62	70	85
18	33	58	90	54	66	74	83	103

TABLE 7.3 Typical Application Data for Silicon-Carbide Intermediate-Class, Nonlinear Resistor Valve-Type Lightning Arresters with Gaps (Electrical Characteristics)

Arrester rating, kV rms maximum valve-off or maximum reseal rating	Maximum circuit voltage phase to phase, kV rms		Maximum front-of-wave impulse sparkover, kV crest	Maximum 100% impulse sparkover, 1.2 × 50 wave, kV crest	Minimum 60-Hz sparkover, kV rms	Maximum discharge voltage with discharge current, 8 × 20 wave, kV crest				
	Ungrounded neutral 100% arrester	Effectively grounded neutral 80% arrester				1.5 kA	3 kA	5 kA	10 kA	20 kA
3	3	3.75	11	11	4.5	5.2	6	6.6	7.5	8.7
4.5	4.5	5.63	16	15	6.8	7.8	9	9.9	11.3	13.1
6	6	7.50	21	19	9	10.4	11.9	13.2	15	17.4
7.5	7.5	9.38	26	23.5	11.3	13	14.9	16.5	18.8	21.8
9	9	11.25	31	27.5	13.5	15.6	17.9	19.8	22.5	26.1
10	10	12.50	35	31	15	17.5	20.0	22.0	25.0	29.0
12	12	15.00	40	35.5	18	20.8	23.8	26.4	30	34.8
15	15	18.75	50	43.5	22.5	25.9	29.7	32.9	37.5	43.5
18	18	22.50	59	51.5	27	31.1	35.7	39.5	45	52.2
21	21	26.25	68	59	31.5	36.3	41.6	46.1	52.5	60.9
24	24	30.00	78	67	36	41.5	47.6	52.7	60.0	69.6
27	27	33.75	88	75	40.5	46.7	53.5	59.2	67.5	78.3
30	30	37.50	97	81	45	51.8	59.4	65.8	75	87
36	36	45.00	116	95	54	62.2	71.3	79	90	104.4
39	39	48.75	126	102	58.5	67.4	77.3	85.5	97.5	113.1
45	45	56.25	144	116	67.5	77.7	89.1	98.7	112.5	130.5
48	48	60.0	154	123	72	82.9	95.1	105.3	120	139.2
60	60	75.00	190	153	90	103.6	119	131.2	150	174
72	72	90.00	228	180	108	124.3	142.6	158	180	208.8
78	78	97.50	245	195	117	134.7	154.5	171	185	226.2
84	84	105.00	262	209	126	145	166.4	184.2	210	243.6
90	90	112.50	282	223	135	155.4	178.2	197.3	225	261
96	96	120.00	300	236	144	165.7	190.1	210.5	240	278.4
108	108	135.00	335	263	162	186.5	213.9	237	270	313.1
120	120	150.00	370	290	180	207.2	238	263	300	347.9

TABLE 7.4 Typical Application Data for Silicon-Carbide Station-Class, Nonlinear Resistor Valve-Type Lightning Arresters with Gaps (Electrical Characteristics)

Arrester rating, maximum permissible line-to-ground kV rms (maximum valve-off or maximum reseal rating)	Minimum 60-Hz sparkover, kV rms	Maximum switching surge sparkover, kV crest	Maximum 100% 1½ × 40 impulse sparkover, kV crest	Maximum impulse sparkover, ASA front-of-wave kV crest	Maximum discharge voltage for discharge currents of 8 × 20 microsecond waveshape with the following crest amplitudes					
					1500 A, kV crest	3000 A, kV crest	5000 A, kV crest	10,000 A, kV crest	20,000 A, kV crest	40,000 A, kV crest
3	5	8	8	12	7	8	8.5	9	10	11.5
6	11	17	17	24	15	16	17	19	20	23
9	16	25	24	35	21	23	24	26	28	31.5
12	22	34	32	45	28	30	32	35	38	42
15	27	42	40	55	35	38	40	44	47	52.5
21	36	56	55	72	47	52	55	60	65	73
24	45	70	65	90	56	61	65	71	76	84
30	54	85	80	105	69	75	80	87	94	105
36	66	104	96	125	83	90	96	105	113	126
39	72	113	104	130	90	98	104	114	123	137
48	90	141	130	155	112	122	130	142	153	168
60	108	169	160	190	137	150	160	174	189	210
72	132	206	195	230	167	184	195	212	230	252
90	160	242	228	271	206	226	240	262	283	315
96	175	257	247	294	222	242	258	280	304	336
108	195	294	266	332	244	264	282	316	333	378
120	220	323	304	370	275	301	320	350	378	420

Metal-oxide valve arresters are metal-oxide surge arresters, constructed with zinc-oxide valve blocks, that do not need gaps (Tables 7.5, 7.6, and Fig. 7.2). The zinc-oxide valve blocks in an arrester effectively insulate the electrical conductors from ground at normal electric operating voltages, limiting the leakage current flow through an arrester to a small value. When a surge develops on the electric system protected by metal-oxide surge arresters, the arrester valve blocks conduct large values of current to ground, limiting the $I \times R$ voltage drop across the arrester. If the connections to the arrester have a low impedance, the stress on the electric system is kept below the basic insulation level (BIL) or critical insulation level of the equipment. The zinc-oxide valve blocks or disks start conducting sharply at a precise voltage level and cease to conduct when the voltage falls below the same level. The arrester zinc-oxide valve blocks have the characteristics of a nonlinear resistance.

The normal-duty distribution arrester ANSI C62.11 design test requirements (Table 7.5) are 9a) low current, long duration; 20 surges with 2000-μs duration and 75-A magnitude, 9b) high current, short duration; two surges with 4-/10-μs duration and 65-kA magnitude, and 9c) duty cycle, 22 5-kA discharges.

The heavy-duty distribution arrester ANSI C62.11 design test requirements (Table 7.6) are 9(a) low current, long duration; 20 surges with 2000-μs duration and 250-A magnitude, 9(b) high current, short duration; two surges with 4-/10-μs duration and 100-kA magnitude, and 9(c) duty cycle, 20 10-kA discharges and two 40-kA discharges.

Tables 7.7 and 7.8 show other arrester characteristics.

Surge Arrester Selection Criteria

- Maximum phase-to-ground system voltage
- Maximum discharge voltage
- Maximum continuous operating voltage (MCOV)
- Basic impulse level of the equipment to be protected

Insulation protective margin equation

$$Protective\ margin = \left[\left(\frac{Insulation\ withstand}{V_{max}} \right) - 1 \right] \times 100$$

The equation is taken from ANSI C62.2, *Application Guide for Surge Arresters*. The minimum recommended margin is 20%.

TABLE 7.5 Normal-Duty Polymer-Housed Metal-Oxide Valve Distribution-Class Surge Arrester Electrical Characteristics

Rated voltage, kV	MCOV, kV	Unit catalog number	0.5 μs maximum 5-kA maximum IR-kV	500-A switching surge maximum IR-kV	8/20 Maximum discharge voltage (kV)					
					1.5 kA	3 kA	5 kA	10 kA	20 kA	40 kA
3	2.55	217253	12.5	8.5	9.8	10.3	11.0	12.3	14.3	18.5
6	5.1	217255	25.0	17.0	19.5	20.5	22.0	24.5	28.5	37.0
9	7.65	217258	33.5	23.0	26.0	28.0	30.0	33.0	39.0	50.5
10	8.4	217259	36.0	24.0	27.0	29.5	31.5	36.0	41.5	53.0
12	10.2	217260	50.0	34.0	39.0	41.0	44.0	49.0	57.0	74.0
15	12.7	217263	58.5	40.0	45.5	48.5	52.0	57.5	67.5	87.5
18	15.3	217265	67.0	46.0	52.0	56.0	60.0	66.0	78.0	101.0
21	17.0	217267	73.0	49.0	55.0	60.0	64.0	73.0	84.0	107.0
24	19.5	217270	92.0	63.0	71.5	76.5	82.0	90.5	106.5	138.0
27	22.0	217272	100.5	69.0	78.0	84.0	90.0	99.0	117.0	151.5
30	24.4	217274	108.0	72.0	81.0	88.5	94.5	108.0	124.5	159.0
36	29.0	217279	134.0	92.0	104.0	112.0	120.0	132.0	156.0	202.0

Courtesy of Ohio Brass Co.

TABLE 7.6 Heavy-Duty Polymer-Housed Metal-Oxide Valve Distribution-Class Surge Arrester Electrical Characteristics

Rated voltage, kV	MCOV, kV	Unit catalog number	0.5 µs 10-kA maximum IR-kV	500-A switching surge maximum IR-kV	8/20 Maximum discharge voltage (kV)					
					1.5 kA	3 kA	5 kA	10 kA	20 kA	40 kA
3	2.55	217602	12.5	8.0	9.5	10.0	10.5	11.0	13.0	15.3
6	5.1	217605	25.0	16.0	19.0	20.0	21.0	22.0	26.0	30.5
9	7.65	217608	34.0	22.5	24.5	26.0	27.5	30.0	35.0	41.0
10	8.4	217609	36.5	23.5	26.0	28.0	29.5	32.0	37.5	43.5
12	10.2	213610	43.5	28.2	38.0	32.9	34.8	38.5	43.8	51.5
15	12.7	213613	54.2	35.0	38.4	41.0	43.4	48.0	54.6	64.2
18	15.3	213615	65.0	42.1	46.0	49.1	52.0	57.5	65.4	76.9
21	17.0	213617	69.5	44.9	49.2	52.5	55.7	61.5	69.9	82.2
24	19.5	213620	87.0	56.4	61.6	65.8	69.6	77.0	87.6	103.0
27	22.0	213622	97.7	63.2	69.2	73.9	78.2	86.5	98.4	115.7
30	24.4	213624	108.4	70.0	76.8	82.0	86.8	96.0	109.2	128.4
36	29.0	213629	130.0	84.2	92.0	98.2	104.0	115.0	130.8	153.8

Courtesy of Ohio Brass Co.

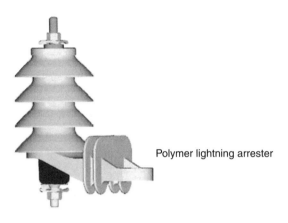

Polymer lightning arrester

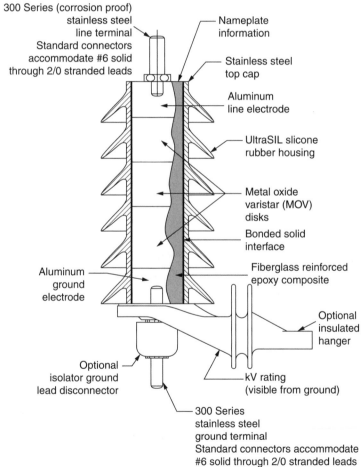

300 Series (corrosion proof)
stainless steel
line terminal
Standard connectors
accommodate #6 solid
through 2/0 stranded leads

Nameplate
information

Stainless steel
top cap

Aluminum
line electrode

UltraSIL slicone
rubber housing

Metal oxide
varistar (MOV)
disks

Bonded solid
interface

Aluminum
ground
electrode

Fiberglass reinforced
epoxy composite

Optional
insulated
hanger

Optional
isolator ground
lead disconnector

kV rating
(visible from ground)

300 Series
stainless steel
ground terminal
Standard connectors accommodate
#6 solid through 2/0 stranded leads

Figure 7.2 Cutaway illustration of UltraSIL housed VariSTAR distribution arrester. (*Courtesy of Coper Power Systems.*)

TABLE 7.7 Riser Pole Polymer-Housed Metal-Oxide Valve Distribution-Class Surge-Arrester Electrical Characteristics

Duty-cycle voltage rating, kV	MCOV, kV	Catalog number	0.5-μs 10-kA maximum IR-kV	500-A switching surge maximum IR-kV	8/20 Maximum discharge voltage, kV					
					1.5 kA	3 kA	5 kA	10 kA	20 kA	40 kA
3	2.55	218603	8.7	5.8	6.5	7.0	7.4	8.1	9.0	10.6
6	5.1	218605	17.4	11.7	13.0	14.0	14.7	16.2	18.1	21.1
9	7.65	218608	25.7	17.5	19.3	21.0	21.9	24.0	27.0	31.6
10	8.4	218609	28.5	19.2	21.2	23.0	24.0	26.5	29.8	34.8
12	10.2	218610	34.8	23.3	25.9	28.0	29.4	32.3	36.2	42.2
15	12.7	218613	43.1	29.1	32.3	35.0	36.6	40.2	45.1	52.7
18	15.3	218615	51.4	34.9	38.6	41.9	43.8	48.0	54.0	63.2
21	17	218617	57.6	38.7	42.8	46.4	48.6	53.6	60.2	70.5
24	19.5	218620	68.8	46.6	51.6	55.9	58.5	64.2	72.1	84.3
27	22	218622	77.1	52.4	57.9	62.9	65.7	72.0	81.0	94.8
30	24.4	218624	85.5	57.6	63.5	69.0	72.0	79.5	89.4	104.4
36	29	218629	102.8	69.8	77.2	83.8	87.6	96.0	108.0	126.6

Courtesy of Ohio Brass Co.

TABLE 7.8   Commonly Applied Voltage Ratings of Arresters

System voltage (kV rms)		Recommended arrester rating per IEEE C 62.22 (kV rms)		
Nominal	Maximum	Four-wire wye multigrounded neutral	Three-wire wye solidly grounded neutral	Delta and ungrounded wye
2.4	2.54	–	–	3
4.16Y/2.4	4.4Y/2.54	3	6	6
4.16	4.4	–	–	6
4.8	5.08	–	–	6
6.9	7.26	–	–	9
8.32Y/4.8	8.8Y/5.08	6	9	–
12.0Y/6.93	12.7Y/7.33	9	12	–
12.47Y/7.2	13.2Y/7.62	9	15	–
13.2Y/7.62	13.97Y/8.07	10	15	–
13.8Y/7.97	14.52Y/8.38	10	15	–
13.8	14.52	–	–	18
20.78Y/12.0	22Y/12.7	15	21	–
22.86Y/13.2	24.2Y/13.87	18	24	–
23	24.34	–	–	30
24.94Y/14.4	26.4Y/15.24	18	27	–
27.6Y/15.93	29.3Y/16.89	21	30	–
34.5Y/19.92	36.5Y/21.08	27	36	–
46Y/26.6	48.3Y/28	36	–	–

Courtesy of Cooper Power Systems.

Example: Determine the margin of protection for both the normal-duty and heavy-duty MOV arrester on a 15-kV class, 13,200/7620-volt, four-wire, overhead system utilizing the 20-kA 8/20-kV discharge voltage.

Solution: From Table 7.1, the 15-kV equipment class BIL is 95, from Table 7.8, a 10-kV rated arrester is utilized. The corresponding maximum discharge voltage is 41.5 for the normal-duty arrester (Table 7.5) and 37.5 for the heavy-duty arrester (Table 7.6):

$$Protective\ margin = \left[\left(\frac{Insulation\ withstand}{V_{max}} \right) - 1 \right] \times 100$$

$$Normal\text{-}duty\ arrester\ protective\ margin = [(95/41.5) - 1] \times 100 = 129\%$$

$$Heavy\text{-}duty\ arrester\ protective\ margin = [(95/37.5) - 1] \times 100 = 153\%$$

Not included are the effects of faster rates of current rise on the discharge voltage, reduced insulation levels because of various factors, and line and ground leads.

Figure 7.3 shows the components comprising the dead-front elbow arrester. Table 7.9 shows the characteristics of dead-front elbow arresters.

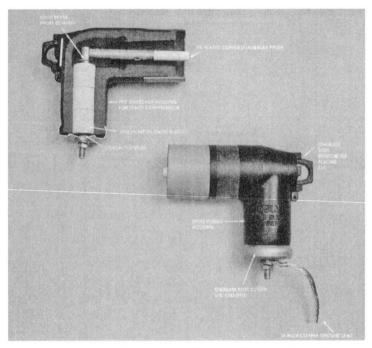

Figure 7.3   Dead-front elbow arrester. (*Courtesy MacLean Power Systems.*)

Example: Determine the margin of protection for a 15-kV underground system utilizing only a riser pole arrester and then utilizing both a riser-pole arrester and an elbow arrester at the open point. Assume the arrester lead length to be 5 ft and utilize the 20-kA 8/20-kV discharge voltage.

Solution: $Protective\ margin = \left[\left(\dfrac{Insulation\ withstand}{V_{max}}\right) - 1\right] \times 100$

Protective margin calculation: arrester only at the riser pole. In the case of utilizing only an arrester at the riser pole, V_{max} consists of both the arrester discharge voltage and the arrester lead length calculated at 1.6 kV per foot. V_{max} is doubled because of the reflection of the wave at the open point in the underground system:

$V_{max} = 2 \times (Discharge\ voltage\ of\ the\ arrester$

$+ \ Voltage\ drop\ caused\ by\ arrester\ lead\ length)$

From Table 7.7, the 10-kV, 20-kA, 8/20 maximum discharge voltage is 29.8. The voltage drop caused by a 5-foot arrester lead length is 5×1.6 or 8.0 kV:

TABLE 7.9   Dead-Front Elbow Arrester Electrical Characteristics

Voltage rating (kV-RMS)	MCOV (kV-RMS)	Equivalent F.O.W.* (kV-CREST)	Maximum discharge voltage (kV-crest) using an 8/20-μs current wave				
			1.5 kA	3 kA	5 kA	10 kA	20 kA
3	2.55	12.9	10.6	11.2	11.7	13.0	15.7
6	5.1	25.8	21.3	22.3	23.5	25.9	31.5
10	8.4	38.7	31.9	33.5	35.2	38.9	47.2
12	10.2	51.6	42.5	44.8	46.9	52.0	62.9
18	15.3	71.1	58.5	61.4	64.6	71.3	86.2
24	19.5	96.8	79.7	84.0	87.9	97.5	117.9
27	22.0	110.0	90.0	94.9	99.7	110.1	133.7

Courtesy MacLean Power Systems.
*The equivalent front of wave is the maximum discharge voltage for a 5-kA impulse current wave, which produces a voltage wave cresting in 0.5 μs.

$$V_{\max} = 2 \times (29.8 + 8.0) = 75.6 \, \text{kV}$$

$$Protective \; margin = \left[\left(\frac{Insulation \; withstand}{V_{\max}} \right) - 1 \right] \times 100$$

$$Protective \; margin = [(95/75.6) - 1] \times 100 = 25.7\%$$

Protective margin calculation: arrester at both the riser pole and the open point. In the case of utilizing an arrester at the riser pole and at the open point of the underground system, $V_{\max}$ consists of the arrester discharge voltage, the arrester lead length calculated at 1.6 kV per foot and some percentage of the front of wave voltage of the open point arrester: (The percentage is typically between 15 and 40%. For this calculation, use a value of 30%.)

$$V_{\max} = (Discharge \; voltage \; of \; the \; arrester$$

$$+ \; Arrestor \; lead \; length \; voltage \; drop$$

$$+ \; 0.3 \times Open \; point \; fow \; voltage)$$

From Table 7.7, the 10-kV, 20-kV, 8/20 maximum discharge voltage is 29.8. The voltage drop from a 5-foot lead length is 5 × 1.6 or 8.0 kV. Finally from Table 7.9, the open point f-o-w voltage of a 10-kV arrester is 38.7. For purposes of the calculation, 30% of 38.7 equals 11.61:

$$V_{\max} = (29.8 + 8.0 + 11.61) = 49.41 \, \text{kV}$$

$$Protective \; margin = \left[\left(\frac{Insulation \; withstand}{V_{\max}} \right) - 1 \right] \times 100$$

$$Protective \; margin = [(95/49.41) - 1] \times 100 = 92.3\%$$

NOTES

8

Fuses

Definitions

Fuses Protection devices connected into circuits to open the circuit and de-energize the apparatus to prevent or limit damage due to an overload or short circuit.

Fuse element An elemental metal (such as silver, tin, lead, copper) or an alloy (such as tin-lead) metal that will melt at a predetermined current maintained for a predetermined length of time.

Blown fuse A fuse that the fuse element has melted or vaporized because of excessive current.

Distribution cutout This provides a high-voltage mounting for the fuse element used to protect the distribution system or the equipment connected to it.

Fuse Links

Figures 8.1 through 8.3 show cutout views of fuse links.

Standard Fuse Link Ratings

Preferred ratings 6, 10, 15, 25, 40, 65, 100, 140, 200
Non-preferred ratings 8, 12, 20, 30, 50, 80

Distribution Cutouts

Figures 8.4 through 8.6 show fuse link cutouts, and Fig. 8.7 is a diagram of fuse operation.

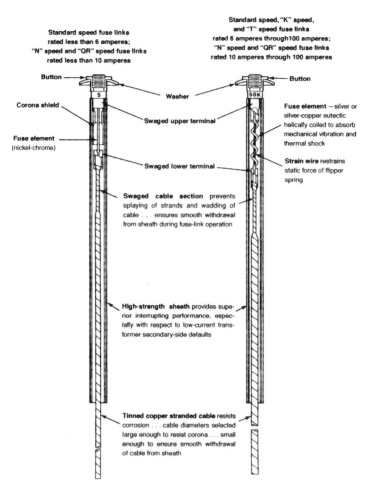

Figure 8.1 A typical primary cutout fuse link showing component parts: button, upper terminal, fuse element, lower terminal, cable (leader), and sheath. The helically coiled fuse element is typical of silver-element fuse links. (*Courtesy of S&C Electric Co.*)

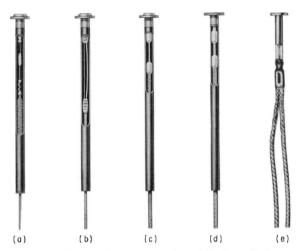

(a)   (b)   (c)   (d)   (e)

Figure 8.2 These primary cutout fuse links have different fusible elements. (a) a high-surge dual element (1 to 8 amps); (b) a wire element (5 to 20 amps); (c) a die-cast tin element (25 to 50 amps); (d) a die-cast tin element (65 to 100 amps); and (e) a formed strip element (140 to 200 amps).

Pull ring

Arc–confining tube

Fusible element

Pull ring

Figure 8.3 The open primary cutout fuse link is provided with pull rings for handling with hot-line tools. Hooks are inserted in rings for placement in the cutout. The cutaway view shows fusible element inside arc-confining tube.

Enclosed cutouts

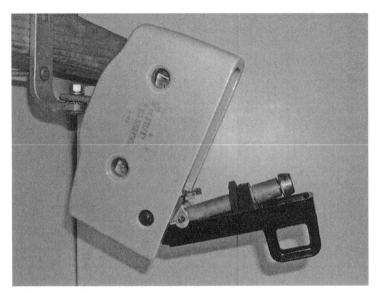

Figure 8.4   Vertical dropout fuse cutout indicating that fuse link has melted. *(Courtesy of ABB Power T&D Company Inc.)*

Open distribution cutouts

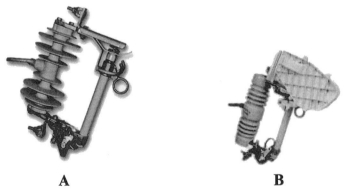

A   B

Figure 8.5   (A) Distribution cutout with expulsion fuse mounted. The fuse is in the closed position, completing circuit between top and bottom terminals. (B) Open distribution cutout with arc chute designed for load break operation. *(Courtesy of ABB Power T&D Company Inc.)*

Open-link primary cutout

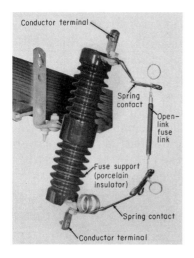

Figure 8.6 Open-link primary cutout showing spring terminal contacts and fuse link enclosed in an arc-confining tube. During fault clearing, the spring contacts provide link separation and arc stretching. (*Courtesy of S&C Electric Co.*)

Fuse operation diagram

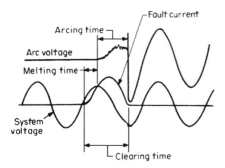

Figure 8.7 Diagram of voltages, current, and timing reference recorded with an oscilloscope graph to show fuse operation.

Time-Current Characteristic Curves

Figures 8.8 through 8.10 show time-current characteristic (TCC) curves for various fuse links.

Protective Fuse to Fuse Overcurrent Coordination

Figure 8.11 shows a simple diagram of a distribution system with two fuses in series that require a fuse to fuse overcurrent protection scheme. Figure 8.12 shows the TCC curve between the protecting downstream fuse and the protected upstream fuse. The rule of thumb is that 75 percent of the minimum melting-time curve of the protected fuse must not overlap the maximum clearing-time curve of the protecting fuse. In some cases, overlapping of the curves is unavoidable (Fig. 8.13). As long as the fault current does not exceed specified ampere values, coordination will occur (Table 8.1).

Transformer Fusing Schedules

Table 8.2 shows the suggested fusing schedule for various sizes of transformers.

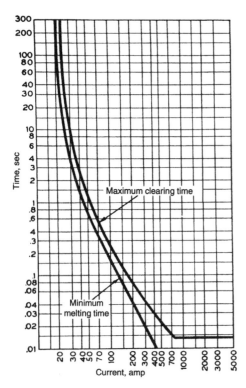

Figure 8.8 Typical time-current characteristic curves (TCC) for 10K fuse link. The minimum melting-time curve and the maximum clearing-time curve are plotted for the 10K fuse link. Fault currents in units of amperes are plotted along the horizontal axes, and time to blow in units of time are plotted along the vertical axes. TCC curves are plotted on logarithmic scale paper. (*Courtesy of S&C Electric Co.*)

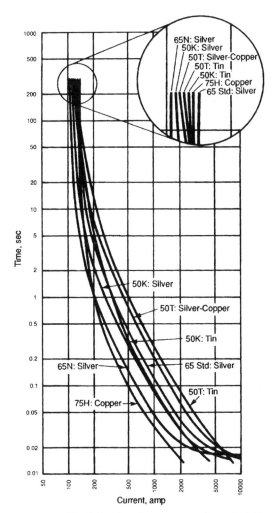

Figure 8.9   Total clearing time-current characteristic curves for fuse links recommended for a 1200-kvar grounded wye-connected capacitor bank rated 13.8-kV three-phase, with two 7.97-kV, 200-kvar capacitor units per phase. (*Courtesy of S&C Electric Co.*)

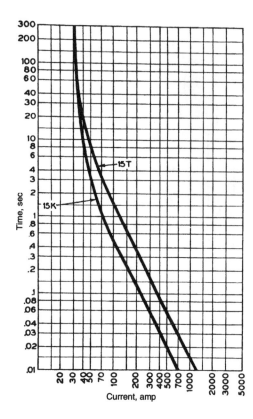

Figure 8.10 The curves show melting times of fast (K) links and slow (T) links having the same 15-amps rating.

Fuse to Fuse Overcurrent Coordination

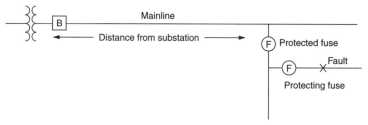

Figure 8.11 For the two fuses depicted; the protecting fuse must melt before the protected fuse in the situation of a fault at location X.

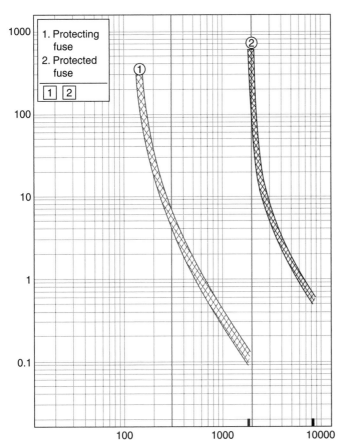

Figure 8.12 The maximum clearing-time of the protecting fuse (curve 1) occurs prior to the minimum melting-time of the protected fuse (curve 2).

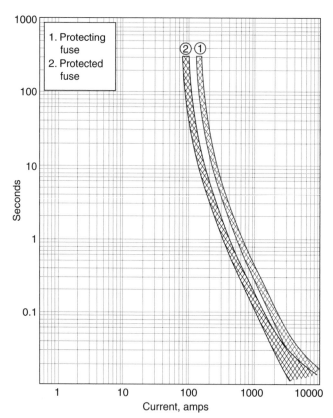

Figure 8.13   There is an overlap of the two fuse TCC curves. The fuses will coordinate as long as the fault current does not exceed values listed in Table 8.1. The maximum clearing-time of the 40T protecting fuse (curve 1) occurs prior to the minimum melting-time of the 65T protected fuse (curve 2) in the lower portion of the curve overlap.

TABLE 8.1 EEI-NEMA Type T Fuse Links

Protecting fuse-link rating– amperes	Protected link rating–amperes													
	8T	10T	12T	15T	20T	25T	30T	40T	50T	65T	80T	100T	140T	200T
	Maximum fault-current protection provided by protecting link–amperes													
6T		350	680	920	1200	1500	2000	2540	3200	4100	5000	6100	9700	15200
8T			375	800	1200	1500	2000	2540	3200	4100	5000	6100	9700	15200
10T				530	1100	1500	2000	2540	3200	4100	5000	6100	9700	15200
12T					680	1280	2000	2540	3200	4100	5000	6100	9700	15200
15T						730	1700	2500	3200	4100	5000	6100	9700	15200
20T							990	2100	3200	4100	5000	6100	9700	15200
25T								1400	2600	4100	5000	6100	9700	15200
30T									1500	3100	5000	6100	9700	15200
40T										1700	3800	6100	9700	15200
50T											1750	4400	9700	15200
65T												2200	9700	15200
80T													7200	15200
100T													4000	13800
140T														7500

This table shows maximum values of fault currents at which EEI-NEMA Type T fuse links will coordinate with each other. The table is based on maximum clearing-time curves for protecting links and 75 percent of minimum melting-time curves for protected links. (*Courtesy of Cooper Power System, Inc.*)

TABLE 8.2 Suggested Transformer Fusing Schedule (Protection Between 200 and 300 percent Rated Load)* (Link sizes are for EEI-NEMA K or T fuse links except for the H links noted.)

Transformer size, kVA	2,400 Δ				4,160Y/2,400		4,800 Δ				8,320Y/4,800	
	Figures 1 and 2		Figure 3		Figures 4, 5, and 6		Figures 1 and 2		Figure 3		Figures 4, 5, and 6	
	Rated amp	Link rating	Rated amp	Link rating	Rated amp	Link rating	Rated amp	Link rating	Rated amp	Link rating	Rated amp	Link rating
3	1.25	2H	2.16	3H	1.25	2H	0.625	1H†	1.08	1H	0.625	1H†
5	2.08	3H	3.61	5H	2.08	3H	1.042	1H	1.805	3H	1.042	1H
10	4.17	6	7.22	10	4.17	6	2.083	3H	3.61	5H	2.083	3H
15	6.25	8	10.8	12	6.25	8	3.125	5H	5.42	6	3.125	5H
25	10.42	12	18.05	25	10.42	12	5.21	6	9.01	12	5.21	6
37.5	15.63	20	27.05	30	15.63	20	7.81	10	13.5	15	7.81	10
50	20.8	25	36.1	50	20.8	25	10.42	12	18.05	25	10.42	12
75	31.25	40	54.2	65	31.25	40	15.63	20	27.05	30	15.63	20
100	41.67	50	72.2	80	41.67	50	20.83	25	36.1	50	20.83	25
167	69.4	80	119.0	140	69.4	80	34.7	40	60.1	80	34.7	40
250	104.2	140	180.5	200	104.2	140	52.1	60	90.1	100	52.1	60
333	138.8	140	238.0		138.8	140	69.4	80	120.1	140	69.4	80
500	208.3	200	361.0		208.3	200	104.2	140	180.5	200	104.2	140

Transformer size, kVA	7,200 Δ				12,470Y/7,200		13,200Y/7,620				12,000 Δ	
	Figures 1 and 2		Figure 3		Figures 4, 5, and 6		Figures 4, 5, and 6		Figures 1 and 2		Figure 3	
	Rated amp	Link rating	Rated amp	Link rating	Rated amp	Link rating	Rated amp	Link rating	Rated amp	Link rating	Rated amp	Link rating
3	0.416	1H†	0.722	1H†	0.416	1H†	0.394	1H†	0.250	1H†	0.432	1H†
5	0.694	1H†	1.201	1H	0.694	1H†	0.656	1H†	0.417	1H†	0.722	1H†
10	1.389	2H	2.4	5H	1.389	2H	1.312	2H	0.833	1H†	1.44	2H
15	2.083	3H	3.61	5H	2.083	3H	1.97	3H	1.25	1H	2.16	3H
25	3.47	5H	5.94	8	3.47	5H	3.28	5H	2.083	3H	3.61	5H
37.5	5.21	6	9.01	12	5.21	6	4.92	6	3.125	5H	5.42	6
50	6.49	8	12.01	15	6.94	8	6.56	8	4.17	6	7.22	10
75	10.42	12	18.05	25	10.42	12	9.84	12	6.25	8	10.8	12
100	13.89	15	24.0	30	13.89	15	13.12	15	8.33	10	14.44	15
167	23.2	30	40.1	50	23.2	30	21.8	25	13.87	15	23.8	30
250	34.73	40	59.4	80	34.73	40	32.8	40	20.83	25	36.1	50
333	46.3	50	80.2	100	46.3	50	43.7	50	27.75	30	47.5	65
500	69.4	80	120.1	140	69.4	80	65.6	80	41.67	50	72.2	80

(Continued)

TABLE 8.2 Suggested Transformer Fusing Schedule (Protection Between 200 and 300 percent Rated Load)* (Link sizes are for EEI-NEMA K or T fuse links except for the H links noted.) (Continued)

Transformer size, kVA	13,200 Δ				14,400 Δ				24,900Y/14,400	
	Figures 1 and 2		Figure 3		Figures 1 and 2		Figure 3		Figures 4, 5, and 6	
	Rated amp	Link rating	Rated amp	Link rating	Rated amp	Link rating	Rated amp	Link rating	Rated amp	Link rating
3	0.227	1H†	0.394	1H†	0.208	1H†	0.361	1H†	0.208	1H†
5	0.379	1H†	0.656	1H†	0.347	1H†	0.594	1H†	0.374	1H†
10	0.757	1H†	1.312	2H	0.694	1H†	1.20	2H	0.694	1H†
15	1.14	1H	1.97	3H	1.04	1H	1.80	3H	1.04	1H
25	1.89	3H	3.28	5H	1.74	2H	3.01	5H	1.74	2H
37.5	2.84	5H	4.92	6	2.61	3H	4.52	6	2.61	3H
50	3.79	6	6.56	8	3.47	5H	5.94	8	3.47	5H
75	5.68	8	9.84	12	5.21	6	9.01	12	5.21	6
100	7.57	8	13.12	15	6.94	8	12.01	15	6.94	8
167	12.62	15	21.8	25	11.6	12	20.1	25	11.6	12
250	18.94	25	32.8	40	17.4	20	30.1	40	17.4	20
333	25.23	30	43.7	50	23.1	30	40.0	50	23.1	30
500	37.88	50	65.6	80	34.7	40	60.0	80	34.7	40

*Reprinted with permission from "Distribution System Protection and Apparatus Coordination," published by Line Material Industries.
†Since this is the smallest link available and does not protect for 300 percent load, secondary protection is desirable.

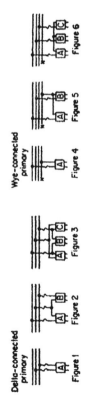

Delta-connected primary

Figure 1   Figure 2   Figure 3

Wye-connected primary

Figure 4   Figure 5   Figure 6

NOTES

NOTES

Chapter

9

Inspection Checklists for Substations, Transmission Lines, and Distribution Lines

Substation Inspection

Substations are inspected to verify all equipment is functioning properly, to proactively diagnose conditions that may damage the equipment, to minimize service interruptions, and to record equipment operations. Many of the devices that operate within the substation may be monitored on a daily basis, remotely, by the supervisory control and data acquisition (SCADA) system. SCADA allows for the monitoring of system bus voltages, power system amp and var flows, and equipment alarms.

On a monthly basis, inspection of the following items are reviewed:

- Power transformers (including general transformer appearance, bushings free of contamination, no oil leaks, auxiliary cooling system, no bird nests, and doors closed)
- Circuit breakers (including general breaker appearance, bushings free of contamination, no oil leaks, doors locked and working)
- Substation insulators
- Steel superstructures
- Bus support insulators
- Suspension insulators
- Circuit switchers

- Disconnect switches
- Coupling capacitors
- Capacitors
- Cable potheads
- Lightning arresters
- Metal-clad switchgear
- Relays
- Communication equipment
- Back-up battery systems
- Control house
- Locks on switches, enclosures, and gates
- Fences, gates, and warning signs (including washouts)
- Grounds and the grounding system (including broken, loose, or exposed wires and exposed ground rods)
- Weeds, paint, and gravel

Additional inspections or patrols should be carried out following significant weather-related events that might affect the equipment.

Specific equipment tests on the equipment are performed based upon frequency of operation. Some of these tests are:

- Transformer gas-in-oil analysis
- Oil dielectric tests
- Relay tests
- Infrared tests
- Voltage regulation equipment tests
- Predictive maintenance tests of load tap changer motor-control circuitry, and of breaker operator mechanisms
- Battery and battery-charger tests

Transmission-Line Inspection

Initial inspection

All transmission lines should be inspected after construction is completed before energizing the line. Linemen should climb each structure and check the following:

- Conductor condition.
- Conductor sag and clearance to ground, trees, and structures.

- Insulator conditions.

- Line hardware for roughness and tightness. Excess inhibitor found should be removed from conductors to prevent corona discharges.

- Structure vibration and alignment.

- Guys for anchors that are pulling out, guy-wire conditions, or missing guy guards.

- Ground-wire connections and conditions.

- Ground resistance at each structure.

- Structure footings for washouts or damage.

- Aircraft warning lights operating properly.

Aerial inspections

- Damaged insulators

- Broken or damaged conductors

- Leaning or downed structures

- Tree clearance problems

Walking inspections

- Repeat the initial climbing inspection and aerial inspection checklist

- Measure radio or TV interference voltages to detect corona discharges and defective insulators

Distribution-Line Inspection Checklist

Common defects or conditions, that the inspector should observe, record, and recommend to be repaired, are as follows:

1. Poles
   - Broken or damaged poles needing replacement
   - Leaning poles in unstable soil
   - Poles in hazardous locations

2. Guys
   - Slack, broken, or damaged guys
   - Guy too close to primary conductors or equipment
   - Guy insulator not installed as required

3. Pole-top assemblies
   - Broken, burnt, or damaged pins and/or crossarm requiring replacement
   - Broken skirt of pin-type insulators

4. Line conductors
   - Too much slack in primary conductors
   - Floating or loose conductors
   - Conductors burning in trees
   - Foreign objects on line
   - Insufficient vertical clearance over and/or horizontal clearance from:
     - Other wires
     - Buildings, parks, playgrounds, roads, driveways, loading docks, silos, signs, windmills, and spaces and ways accessible to pedestrians
     - Waterways—sailboat-launching area
     - Railways
     - Airports
     - Swimming pools
5. Equipment installations
   - Equipment leaking oil
   - Arresters blown or operation of ground lead isolation device
   - Blown fuses
   - Switch contacts not properly closed

Distribution-Line Equipment Maintenance

Capacitor maintenance

All switched capacitor banks should be inspected and checked for proper operation once each year prior to the time period they are automatically switched on and off to meet system requirements.

Capacitor-bank oil switches should be maintained on a schedule related with the type of on/off controls installed at each bank. The maximum number of open and close operations between maintenance of the switches should not normally exceed 2500. Experience has shown that the following schedule will normally keep the equipment operating properly:

Type of control	Maintenance schedule, years
Time clock	3
Voltage	3
Dual temperature	5
Temperature only	8
Time clock and temperature	8

Recloser maintenance

Hydraulically controlled oil circuit reclosers should normally be maintained after 50 operations, or every 2 years. Electronically controlled oil reclosers should be maintained after approximately 150 operations, or every 2 years. The hydraulically or electronically controlled oil reclosers can be maintained in the field or taken to the distribution shop. The dielectric strength of the insulating oil should be preserved by filtering the oil until the tests ensure good results. The recloser manufacturer's instructions should be followed to complete the inspection and repairs.

Vacuum reclosers require field testing on a 6-year schedule. An excessive number of recloser operations may make it necessary to reduce the scheduled maintenance time period. High-potential testing of the vacuum interrupter bottles can be completed in the field by the lineman. The electronic controls for the vacuum recloser can be tested in the field by following the manufacturer's instructions.

All line reclosers should be installed with bypass provisions and a means for easy isolation of the equipment for maintenance work.

Pole-mounted switch maintenance

Group-operated or single-pole switches installed in overhead distribution lines should be checked each time they are operated by the lineman for the following:

- Burnt contacts
- Damaged interrupters or arcing horns
- Proper alignment
- Worn parts
- Defective insulators
- Adequate lightning arrester protection
- Proper ground connections
- Loose hardware

Underground distribution circuit maintenance

Circuits for underground distribution originate at

- A substation
- A pad-mounted switchgear
- The riser pole off of the overhead distribution

Riser pole maintenance

The riser pole for underground distribution circuits should be inspected when overhead lines are inspected and maintained. The inspection should include the disconnect switches and/or fused cutouts, the lightning arresters, and the integrity of the arrester ground lead isolator.

Switchgear maintenance

The switchgear for underground electric distribution circuits should be inspected annually. The inspection should start with external items such as condition of the pad or foundation. Pads that are tipped or cracked or that have the dirt washed out from under them should be repaired. The paint on metal surfaces of the gear should be in good condition to prevent rusting. The cabinet doors should be locked and in most cases bolted shut. Cleanliness of the internal areas of the switchgear is important to maintain the dielectric strength of the insulation. Equipment with insulating oil must be checked for moisture or contamination of the oil. Switches, fuses, and other internal equipment should be inspected in accordance with the manufacturer's instructions. Cable terminations, grounds, and other connections to the switchgear equipment must be kept in good condition.

Distribution transformer maintenance

Transformers should be checked for proper loading. Transformer loadings are normally monitored by use of a transformer load–management system. The system takes monthly kilowatthour meter readings, used for billing purposes for all customers served from a transformer, and converts them to a value of kVA load that is compared to the kVA rating of the transformer. Overloaded transformers should be replaced with ones of adequate capacity to prevent the transformer from failing and to prevent fires.

Pole-type transformers should be inspected visually for defects as a part of the overhead line inspection. Pad-mounted transformers are inspected for the items discussed previously for the underground distribution circuit switchgear. Annual inspections are desirable.

Underground cable maintenance

The terminations of underground cables are inspected as a part of the maintenance procedures for riser poles, switchgear, and pad-mounted transformers. The route of underground cable circuits should be walked by the lineman on a 2-year cycle. The route of the cable should be inspected for obstacles that could damage the cables, changes in grade of the ground that could cause the cables to be exposed or damaged, and washouts. Cables that fail while in operation must be located and repaired or replaced. Secondary enclosures for low-voltage cables should be inspected while checking the route of primary cables or while connecting or disconnecting services to customers. The lid should be removed from the enclosure and checked visually for cable damage, rodent problems, and washouts. Required repairs can usually be completed by the lineman or cableman at the time of the inspection.

Underground primary distribution cables with solid-dielectric insulation have experienced a high rate of electrical failure after several years of operation as a result of *electrochemical trees*. The electrochemical treeing is generally caused by the presence of water in the conductor that penetrates the insulation. It has been successfully demonstrated that the life of the cables can be extended by flushing the conductor with dry nitrogen to remove the moisture and then filling the conductor with a pressurized high-dielectric fluid compatible with the insulation. Contractors are available to complete this maintenance for those companies inexperienced in the procedure.

Street light maintenance

The lumen output of any electric lamp decreases with use, and accumulated dirt interferes with light reflection and transmission. Street light installations are designed such that the average footcandle levels recommended by the *ASA Standards for Roadway Lighting* represent the average illumination delivered on the roadway when the illuminating source is at its lowest output and when the luminaire is in its dirtiest condition. The luminaire refractor should be washed and the lamp replaced when the lumen output of the lamp has fallen to approximately 80 percent of its initial value, to maintain the designed footcandle level.

Lamp burnout should be a consideration in the maintenance of a street light system. The *rated lamp life* is defined as the time required for half of the lamps to fail. The mortality curves for the various types of lamps are relatively flat for the initial 60 to 70 percent of rated life. Using a 6000-hour incandescent lamp as an example, only 10 percent of the lamps will fail in the first 4000 hours of burning time, while 40 percent will fail in the next 2000 hours of burning time.

Proper maintenance of street and highway lighting is important to provide a high degree of service quality. Street lights should be inspected on a 10-year schedule. Refractors for enclosed-type street light luminaires should be cleaned and washed at the time of inspection. The luminaires should be checked for proper level in both the transverse and horizontal directions. If the street light is installed on a steep grade, the transverse direction of the luminaire should be leveled parallel to the street grade. Defective or broken refractors or luminaires must be replaced.

Many street light problems and failures are reported by the public, police, and employees of the concern owning the system. Lamps that fail, defective photocells, and ballasts should be replaced promptly to restore the lighting service. Group replacement of mercury-vapor and high-pressure sodium-vapor lamps on a 10-year schedule should minimize the number of lamp failures and maintain light output in accordance with established standards. The group replacement can be coordinated with the inspection, cleaning, and washing schedule to minimize the effort required.

Voltage regulator maintenance

Distribution voltage regulators should be inspected externally by the inspector while patrolling the overhead lines. Problems such as damaged insulators, leaking oil, or unusual operating noise should be corrected by the lineman as soon as practical. Operation of the voltage regulators should be observed by checking the operation counter and reading the voltage with the aid of recording voltmeters at regular intervals set at various locations along the circuit beyond the regulator. The voltage regulator should be removed from service by the lineman and maintained in accordance with the manufacturer's recommendation as determined by the number of operations of the tap changer or time in service. Normally, the voltage regulators can operate several years between maintenance periods. All manufacturer's require that the insulating oil be kept dry and free from carbon to maintain the dielectric strength of the oil.

Maintenance of Equipment Common to Transmission and Distribution Lines

Pole maintenance

All wood poles should be pounded with a hammer by the lineman and visibly inspected before being climbed. A good pole will sound solid when hit with a hammer. A decayed pole will sound hollow or like a drum. The visible inspection should include checking for general condition and position, cracks, shell rot, knots that might weaken the pole, setting depth, soil conditions, hollow spots or woodpecker holes, and burned spots. Poles that reveal possible defects should not be climbed by the linemen until the poles are made safe by repairs or by securing them.

All wood poles should be pressure- or thermal-treated with preservatives when they are new. New treated wood poles should resist decay for 10 to 20 years, depending upon the ambient soil conditions. Scheduled inspection should be based on experience. Distribution poles are often replaced before scheduled maintenance is required because of a change in voltage or replacement due to street-widening projects. When a lineman detects defects in wood poles of a particular type and vintage, similar poles installed at the same time should be inspected and maintained, if necessary. A pole hit with a hammer by the lineman will not always reveal decay.

The Electrical Power Research Institute (EPRI) has developed an instrument that uses nondestructive test methods to predict the life of a pole. The lineman keys in the wood species and the diameter of the pole on the instrument and then strikes the pole a few times with a marble-size steel ball that has been attached to the pole. A microprocessor in the instrument analyzes the sonic wave patterns passing through the wood pole and indicates the current strength of the pole.

If the instrument indicates the pole strength is deficient or if a pole-testing instrument is not available, the ground around the pole can be excavated to a depth of 12 to 18 inches. A probe test can be completed by hand-drilling a 3/8-inch hole below the ground line into the pole butt at a 30° to 45° downward angle to a depth sufficient to reach the center of the pole. The wood shavings are checked for decay. The drilled holes are plugged with a treated 3/8-inch dowel to keep moisture out and to prevent further decay. Poles that have sufficient decay to weaken them should be reinforced or replaced. External decay of poles should be shaved away. Poles in acceptable condition should have a preservative applied at the ground line and be wrapped with a moisture-proof barrier. The wood poles can be treated internally with chemicals to kill early decay before actual voids form. Poles can be strengthened to delay replacement by placing a section of galvanized steel plate adjacent to the pole. The steel plate is driven into the ground and banded to the pole above the ground line.

Wood poles installed near woodpecker's feeding areas are often damaged by the birds. Woodpecker damage can be prevented by wrapping the pole with extruded sheets of high-density polyethylene with a high-gloss finish. The plastic wrap prevents the woodpeckers from perching on the pole. Lineman can climb the wrapped pole by penetrating the wrap with climber's gaffs. If the linemen must do considerable work on the pole, the pole should be rewrapped when the work is completed to prevent sites of gaff penetration of the plastic wrap from becoming perches for the birds.

Wooden transmission-line poles are usually scheduled for inspection after 20 years of service and treated if pole conditions are adequate. Repeat inspections and treatment are normally scheduled at 8- to 10-year intervals. Experience indicates that poles properly treated when new and maintained on an adequate schedule should have an average life of 45 years.

Insulator maintenance

Transmission- and distribution-line inspections will detect damaged insulators. Most insulator damage will result from gun shots, lightning flashovers, contamination flashovers, and wind damage. Broken, cracked, or damaged insulators should be replaced as soon as they are detected. Defective insulators may cause visible corona and interference voltage propagation. Radio interference conditions can be detected by the lineman by using instruments designed for this purpose while completing the scheduled inspections.

Insulators in severely contaminated atmospheres may require frequent cleaning. These insulators can be washed with high-pressure water, using special hoses and hose nozzles. Procedures developed for safe washing of energized insulators must be followed.

Conductor maintenance

Conductors can be damaged by aeolian vibration, galloping, sway oscillation, unbalanced loading, lightning discharges, or short-circuit conditions. Inspections of the transmission and distribution lines by the lineman will reveal the need for conductor repairs.

Abrasion to the outside surfaces of the conductor is visible to the inspector. Abrasion damage is a chafing, impact wear that accompanies relative movement between a loose tie or other conductor hardware and the conductor or armor rods. Abrasion is a surface damage and can be identified by black deposits on the conductor or tie wire.

Fatigue results in the failure of the conductor strands or tie wire. If the damaged conductor is detected before serious damage or complete failure occurs, it can be repaired by the lineman by installing preformed armor rods or a preformed line splice.

Broken conductor tie wires at an insulator discovered by inspection or as a result of radio interference complaints require immediate repairs. The damaged conductor can be repaired even if some of the conductor strands are broken by removing the old tie wires and armor rods, if present, and installing preformed armor rods or preformed line guards and retying the assemblies with preformed distribution ties or wraplock ties.

Transmission and distribution lines where experience indicates the conductors have prolonged periods of vibration should have vibration dampers installed to prevent fatigue failure of the conductor. Preformed spiral vibration dampers will prevent conductor fatigue, inner wire fretting, or scoring of the distribution-line insulator glaze.

Transmission and distribution lines can be tested with Thermovision equipment to locate hot spots on conductors and equipment caused by loose connections or hardware and other defects. Thermovision equipment locates hot spots by using infrared scanning to thermally inspect the lines. Some of the equipment provides a photographic record as well as a visual display of the hot-spot defects. The Thermovision equipment can be operated from a helicopter, from a van-type vehicle, or by an operator carrying the equipment.

NOTES

Chapter

10

Tree Trimming

Industry Standards

ANSI Z133.1 *American National Standard for Tree Care Operations—Pruning, Trimming, Repairing, Maintaining, and Removing and Cutting Brush-Safety Requirements*

ANSI A-300 *American National Standard for Tree Care Operations—Tree, Shrub and Other Woody Plant Maintenance—Standard Practices,* Secretariat—National Arborist Association, Incorporated.

Definitions

Line clearance   Preventive line maintenance to ensure that the utility's service to its customers is not interrupted as the result of interference with conductors or circuit equipment by growing trees. It should be accomplished while maintaining the health and beauty of the trees involved, the goodwill of property owners and public authorities, and the safety of the trimming crew and the public.

Maintenance trimming cycles   The amount of clearance required to prevent tree interference with the wire for a given period of time. Transmission lines are inspected annually. Distribution line clearance cycles vary between 2 and 3 years. The tree species characteristics, such as flexibility, rate of growth, and the tendency to shed limbs, are considered in determining how much should be trimmed.

Danger tree   Trees that could reach within 5 ft of a point underneath the outside conductor in falling in the direction of the wires.

Buffer zone   Trees and shrubs intentionally planted or left directly under or parallel to electric facilities, to minimize the visual impact to the surrounding environment. Low growth tree species and shrubs are required to maintain proper wire clearance.

Natural trimming   Branches are flush-cut at a suitable parent limb back toward the center of the tree. Better tree form and value are maintained, and regrowth direction can be influenced. This method of trimming is sometimes called *drop-crotch trimming* or *lateral trimming*. Large branches should be removed to laterals at least one-third the diameter of the branch being removed. Natural trimming is especially adapted to the topping of large trees, where a great deal of wood must be removed. In natural trimming, almost all cuts are made with a saw, and very little pole pruning work is required. This when, finished, results in a natural-looking tree, even if a large amount of wood has been removed. Natural trimming is also directional trimming, since it tends to guide the growth of the tree away from the wires. Stubbing or pole-clip clearance, on the other hand, tends to promote rapid sucker growth right back into the conductors. The big factor to remember is that natural clearance does work and that two or three trimming cycles done in this manner will bring about an ideal situation for both the utility and the tree owner. Most shade trees lend themselves easily to this type of trimming. Elm, Norway maple, red oak, red maple, sugar maple, silver maple, and European linden are our most common street trees, and these species react especially well to natural clearance methods.

Side trimming   Cutting back or removing the side branches that are threatening the conductors. Side trimming is required where trees are growing adjacent to utility lines. Limbs should be removed at a lateral branch. Unsightly notches in the tree should be avoided if possible. Shortening branches above and below the indented area or balancing the opposite side of the crown will usually improve the appearance of the tree. When trimming, remove all dead branches above the wires, since this dead wood could easily break off and cause an interruption.

Under trimming   Removing limbs beneath the tree crown to allow wires to pass below the tree. To preserve the symmetry of the tree, lower limbs on the opposite side of the tree should also be removed. All cuts should be flush to avoid leaving unsightly stubs. The natural shape of the tree is retained in this type of trimming, and the tree can continue its normal growth. Overhangs are a hazard, however, when a line passes beneath a tree. Overhangs should be removed in accordance with the species of tree, location, and the general policy of the utility that you work for. When trimming, remove all dead branches above the wires, since this dead wood could easily break off and cause an interruption. Many utilities have a removal program set up for trees that overhang important lines.

Through trimming   The removal of branches within the crown to allow lines to pass through the tree. It is best suited for low-voltage secondary service wires, street and security light wires, and cables, although it is often used on primary circuits where there is no other way of trimming the tree. Cuts should be made at crotches to encourage growth away from the lines.

Trimming Branches and Limbs

Figures 10.1 and 10.2 show proper tree-trimming techniques.

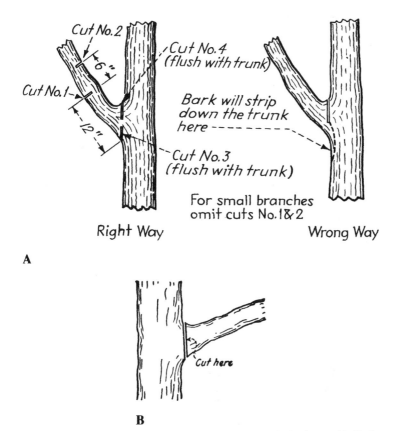

Figure 10.1   (A) Steps in the right and wrong removal of a large side limb. (*Courtesy of REA.*) (B) How a small branch can be removed with a single cut.

Lowering Large Limbs

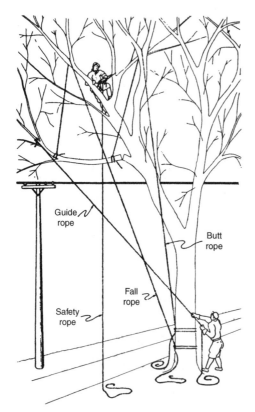

Figure 10.2 Lowering large or hazardous limb with rope tackle. The two supporting ropes are called butt and fall ropes; the third rope is the guide rope. (*Courtesy of Edison Electric Institute.*)

Vehicular Equipment

Figure 10.3 shows a large tree chipper.

Figure 10.3 A wood chipper utilized by the tree crew.

Tools

Figures 10.4 through 10.7 show tools used for tree trimming.

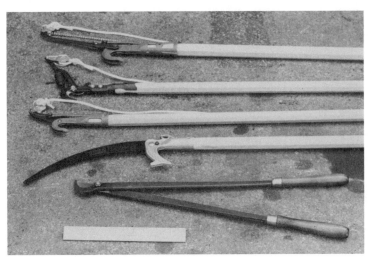

Figure 10.4 Trimming hand tools. Starting at the top, tools are a center-cut bull clip, a side-cut pole clip, a center-cut pole clip, a pole saw, and a pair of loppers. (*Courtesy of Asplundh Tree Expert Co.*)

Figure 10.5 An example of an arborist style hand-saw. (*Courtesy of Asplundh Tree Expert Co.*)

Figure 10.6 Gasoline-powered chain saw utilized by tree crew personnel. (*Courtesy of Stihl.*)

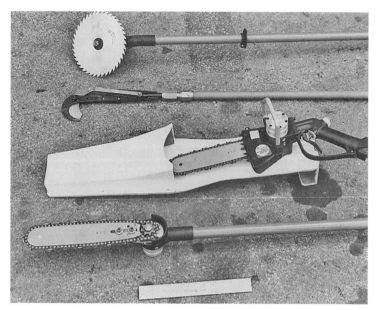

Figure 10.7 Hydraulic tools of a type used from a trim lift bucket. (*Courtesy of Asplundh Tree Expert Co.*)

Trimming Tables

Table 10.1 shows tree trimming clearances and Table 10.2 shows information concerning trimming various types of trees.

TABLE 10.1   Tree Trimming Clearances (in feet)

Clearances	Secondary, 100–600 volts	Primary, 2400–4800 volts	Primary, 7200–13,800 volts
Topping			
Fast growers	6	8	9
Slow growers	4	6	7
Side			
Fast growers	4	6	8
Slow growers	2	4	6
Overhang			
Fast growers	4	8	12*
Slow growers	2	6	12*

*Remove if possible.
SOURCE: Asplundh Tree Expert Co.

TABLE 10.2 Characteristics of Various Trees

Species (common name)	Growth form	Avg. ann. normal growth, in	Avg. ann. sucker growth, in	Mature height, ft	Modulus of rupture (grn.), lb/in.^{2}	Remarks
Ailanthus (tree of heaven)	S	48	60	50		Fast-growing, weak-wooded. Remove if possible.
Alder, Red	U	36	84	120	6,500	Weak-wooded.
Ash	U	18	36	80	9,600	Strong wood, but tends to split. Will not stand heavy topping.
Banyan	S	24	60	70		Rather strong branching. Good crotches for lateral trimming.
Basswood (linden)	S	18	27	75	5,000	Wood not very strong. Tends to split. Watch dead wood.
Beech, American	S	12	30	60	8,600	Strong wood. Mature trees do not stand topping well.
Birch, White	S	21	52	50	6,400	Very sensitive. Intolerant of heavy trimming. Form is spoiled by topping.
Box-elder	H	26	72	50	5,000	Weak-wooded. Suckers grow rapidly. Remove if possible.
Catalpa	S	12		60	5,000	Soft, weak wood. Slow-growing.
Cedar, Eastern Red (juniper)	U	15	15	50	7,000	Host to the Cedar-Apple rust.
Cherry, Wild Black	S	14	24	60	8,000	Found mostly in rural areas. Does not stand heavy topping. Dies back easily. Watch dead wood.
Elm, American	U	26	60	85	7,200	Strong, flexible wood. Avoid tight crotches.
Elm, Chinese	U	40	72	65	5,200	Wood rather brittle. Strips easily.
Eucalyptus (blue gum)	U	24	50	125	11,200	
Ficus	H	36	60	50	5,800	
Fir, Douglas	U	18	21	125	7,600	

(Continued)

TABLE 10.2 Characteristics of Various Trees (Continued)

Species (common name)	Growth form	Avg. ann. normal growth, in	Avg. ann. sucker growth, in	Mature height, ft	Modulus of rupture (grn.), lb/in.^{2}	Remarks
Gum, Sweet	U	12	20	90	6,800	A strong slow-growing tree.
Hackberry	S	18	30	60	6,500	Strong, stringy wood. Similar to elm.
Hickory	U	14	21	65	11,000	Wood very strong and tough. Will not stand heavy topping.
Locust, Black	S	18	80	80	13,800	Normal growth is moderate. Vigorous sucker growth when topped. Remove small trees if possible. Watch thorns.
Locus, Honey	U	22	33	80	10,200	Watch thorns.
Madrona	S	8	24	50	7,600	
Magnolia	U	20	42	90	7,400	Wood splits easily.
Mango	S	18	36	50		
Maple, Big Leaf	S	30	72	60	7,400	Wood tends to split.
Maple, Norway	H	15	35	50	9,400	Wide-spreading as a mature tree. Sucker growth rapid. Can stand heavy trimming.
Maple, Red	S	18	42	75	7,700	Suckers grow rapidly. A rather weak and brittle tree.
Maple, Silver	U	25	65	65	5,800	Mature trees cannot stand heavy topping. Soft wood. Weak branches. Suckers grow very fast. Remove if possible or top under wires.
Maple, Sugar	S	18	40	75	9,400	Not too strong. Rather irregular growth.
Melaleuca (punk tree)	U	36	96	50		Very sturdy trees.
Oak, Australian Silk	U	36	72	50	6,900	Very hard strong wood.
Oak, Black and Red	S	18	30	85	11,900	Strong wood. Topping ruins shape of tree.
Oak, Live	S	30	45	70	8,300	Moderately strong branches.
Oak, Pin	U	24	36	100	8,900	A sturdy and slow-growing tree. Dies back when topped heavily. Sucker growth thin.
Oak, Water	S	30	45	75	8,300	
Oak, White	S	9	18	75		

	Growth form					Notes
Oak, Willow	H	24	40	50	7,400	Strong. Tough. Can be topped safely.
Palm, Coconut	U	60		90		
Palm, Queen	U	36		65		
Palm, Royal	U	48		80		
Palm, Washington	U	36		70		
Pecan	U			140	9,800	Has hard wood. Grows rapidly. Watch weak crotches. Suckers profusely.
Pine, Australian	U	60	144	100		Wood rather strong, but weak branching.
Pine, Shortleaf	U	36	48	80	7,300	Hard brittle wood. Dangerous thorns.
Pithecellobium (ape's earring)	S	24	72	50		
Plane (sycamore)	U	34	72	100	6,500	Rank growth of suckers when topped.
Poinciana, Royal	H	18	36	30		
Poplar, Carolina (cottonwood)	U	52	80	85	5,300	Wood very brittle. Breaks abruptly. Suckers grow at tremendous rate. Remove if possible.
Poplar, Lombardy	U	45	72	60	5,000	Very brittle. Remove if possible. A short-lived tree.
Sassafras		24	36	50	6,000	Does not stand heavy topping. Branches break easily.
Tulip Tree	U	30	52	100	5,400	Wood splits easily.
Walnut. Black	S	20	40	80	9,500	Strong wood. Dies back under heavy topping.
Willow, Black	H	40	70	50	3,800	Very weak and brittle. Breaks easily in storms.
Willow, Weeping	S	48	72	50	3,800	A weak-wooded tree. Remove if possible.

Courtesy of Asplundh Tree Expert Co.
NOTES: Tree growth forms—U—upright, S—spreading, and H—horizontal

NOTES

Rope, Knots, Splices, and Gear

This chapter consists of figures and tables that illustrate various ropes, knots, splices, and gear.

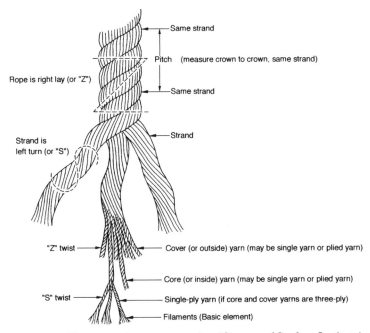

Figure 11.1 Three-strand rope construction. (*Courtesy of Cordage Institute.*)

180

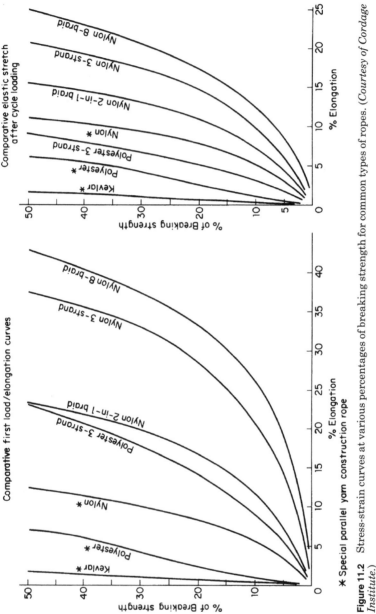

Figure 11.2 Stress-strain curves at various percentages of breaking strength for common types of ropes. (*Courtesy of Cordage Institute.*)

TABLE 11.1 Three-Strand Laid and Eight-Strand Plaited Rope Data

Nominal size, in		Manila				Polypropylene				Composite			
Dia.	Circ.	Linear density, lb./100 ft. (1)	Ultimate strength, lb. (2)	Safety factor	Allowable load, lb. (1)	Linear density, lb./100 ft. (3)	Ultimate strength, lb. (2)	Safety factor	Allowable load, lb. (1)	Linear density, lb./100 ft. (3)	Ultimate strength, lb. (2)	Safety factor	Allowable load, lb. (1)
3/16	5/8	1.44	406	←10→	41	.70	720	←10→	72	.94	720	←10→	72
1/4	3/4	1.90	540		54	1.20	1,130		113	1.61	1,130		113
6/16	1	2.76	900		90	1.80	1,710		171	2.48	1,710		171
3/8	1 1/8	3.90	1,200		122	2.80	2,440		244	3.60	2,440		244
7/16	1 1/4	6.00	1,580	←9→	176	3.80	3,160	←9→	352	5.00	3,160	←9→	352
1/2	1 1/2	7.15	2,380		264	4.70	3,780		420	6.50	3,760		440
9/16	1 3/4	9.90	3,100	←8→	388	6.10	4,600	←8→	575	8.00	4,860	←8→	610
5/8	2	12.70	3,760		476	7.50	5,600		700	9.50	5,750		720
3/4	2 1/4	15.90	4,860	←7→	695	10.70	7,630	←7→	1,070	12.50	7,550	←7→	1,080
13/16	2 1/2	18.60	5,850		835	12.70	8,900		1,270	15.20	9,200		1,310
7/8	2 3/4	21.40	6,950		995	15.00	10,400		1,490	18.00	10,800		1,540
1	3	25.60	8,100		1,160	18.00	12,600		1,800	21.80	13,100		1,870
1 1/16	3 1/4	27.80	9,450		1,350	20.40	14,400		2,060	25.60	15,200		2,180
1 1/8	3 1/2	34.20	10,800		1,540	23.80	16,500		2,360	29.00	17,400		2,480
1 1/4	3 3/4	37.60	12,200		1,740	27.00	18,900		2,700	33.40	19,800		2,820
1 5/16	4	45.60	13,500		1,930	30.40	21,200		3,020	35.60	21,200		3,000
1 1/2	4 1/2	57.00	16,700		2,380	38.40	26,800		3,820	43.00	26,800		3,820
1 5/8	5	71.00	20,200		2,870	47.60	32,400		4,620	55.50	32,400		4,620
1 3/4	5 1/2	85.00	23,800		3,400	57.00	38,800		5,550	66.50	38,800		5,550
2	6	102.00	28,000		4,000	67.00	46,800		6,700	78.00	46,800		6,700
2 1/8	6 1/2	119.00	32,400		4,620	80.00	55,000		7,850	92.00	55,000		7,850
2 1/4	7	137.00	37,000		5,300	92.00	62,000		8,850	105.00	62,000		8,850
2 1/2	7 1/2	159.00	41,800		5,950	107.00	72,000		10,300	122.00	72,000		10,300

(Continued)

TABLE 11.1 Three-Strand Laid and Eight-Strand Plaited Rope Data (*Continued*)

Nominal size, in		Manila				Polypropylene				Composite			
Dia.	Circ.	Linear density, lb./100 ft. (1)	Ultimate strength, lb. (2)	Safety factor	Allowable load, lb. (1)	Linear density, lb./100 ft. (3)	Ultimate strength, lb. (2)	Safety factor	Allowable load, lb. (1)	Linear density, lb./100 ft. (3)	Ultimate strength, lb. (2)	Safety factor	Allowable load, lb. (1)
2⅝	8	182.00	46,800		6,700	130.00	81,000		11,600	138.00	81,000		11,600
2⅞	8½	204.00	52,900		7,400	137.00	91,000		13,000	155.00	91,000		13,000
3	9	230.00	57,500	←7→	8,200	153.00	103,000	←7→	14,700	174.00	103,000	←7→	14,700
3¼	10	284.00	67,500		9,950	190.00	123,000		17,600	210.00	123,000		17,600
3½	11	348.00	82,000		11,700	232.00	146,000		20,800	256.00	146,000		20,800
4	12	414.00	94,500		13,500	276.00	171,000		24,400	300.00	171,000		24,400

Polyester				Nylon				Nominal dia. in.
Linear density, lb./100 ft. (3)	Ultimate strength lb. (2)	Safety factor	Allowable load, lb. (1)	Linear density, lb./100 ft. (3)	Ultimate strength, lb. (2)	Safety factor	Allowable load, lb. (1)	
1.20	900	←10→	90	1.00	900	←12→	75	3/16
2.00	1,490		149	1.50	1,490		124	1/4
3.10	2,300		230	2.56	2,300		192	5/16
4.50	3,340	←9→	334	3.50	3,340		278	3/8
6.20	4,500		500	5.00	4,500	←11→	410	7/16
8.00	5,750		640	6.50	5,750		525	1/2
10.20	7,200	←8→	900	8.15	7,200	←10→	720	9/16
15.00	9,000		1,130	10.50	9,350		935	5/8
17.50	11,300	←7→	1,610	14.50	12,800	←9→	1,430	3/4
21.00	14,000		2,000	17.00	15,300		1,700	13/16
25.00	16,200		2,320	20.00	18,000		2,000	7/8

30.40	19,800	2,820	26.40	22,600	2,520	1
34.40	23,000	3,280	29.00	26,000	2,880	1 1/16
40.00	26,600	3,800	34.00	29,800	3,320	1 1/8
46.20	29,800	4,260	40.00	33,800	3,760	1 1/4
52.50	33,800	4,820	45.00	38,800	4,320	1 5/16
67.00	42,200	6,050	55.00	47,800	5,320	1 1/2
82.00	51,500	7,350	66.50	58,500	6,500	1 5/8
98.00	61,000	8,700	83.00	70,000	7,800	1 3/4
118.00	72,000	10,300	95.00	83,000	9,200	2
135.00	83,000	11,900	109.00	95,500	10,600	2 1/8
157.00	95,500	13,800	127.00	113,000	12,600	2 1/4
181.00	110,000	15,700	147.00	126,000	14,000	2 1/2
204.00	123,000	17,600	168.00	146,000	16,200	2 5/8
230.00	139,000	19,900	189.00	162,000	18,000	2 7/8
258.00	157,000	22,400	210.00	180,000	20,000	3
318.00	189,000	←7→ 27,000	264.00	226,000	←9→ 25,200	3 1/4
384.00	228,000	32,600	312.00	270,000	30,000	3 1/2
454.00	270,000	38,600	380.00	324,000	36,000	4

Courtesy of Cordage Institute.

(1) Allowable loads are for rope in good condition with appropriate splices, in noncritical applications, and under normal service conditions. Allowable loads should be exceeded only with expert knowledge of conditions and professional estimates of risk. Allowable loads should be reduced where life, limb, or valuable property are involved, and/or for exceptional service conditions, such as shock loads, sustained loads, etc.

(2) Ultimate strengths are based on tests of new and unused rope in accordance with Cordage Institute methods and are exceeded at a 98 confidence percent level.

(3) Linear density (lb. per 100 ft.) shown is "average." Maximum weight is 5 percent higher.

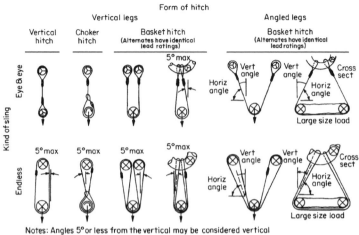

Notes: Angles 5° or less from the vertical may be considered vertical

Explanation of symbols: Minimum diameter of curvature

⊖ Represents a contact surface which shall have a diameter of
curvature at least double the diameter of the rope from which
the sling is made.

⊗ Represents a contact surface which shall have a diameter of
curvature at least 8 times the diameter of the rope.

◔ Represents a load in a choker hitch and illustrates the rotary
force on the load and/or the slippage of the rope in contact
with the load. Diameter of curvature of load surface shall be
at least double the diameter of the rope.

Figure 11.3   Sling configurations.

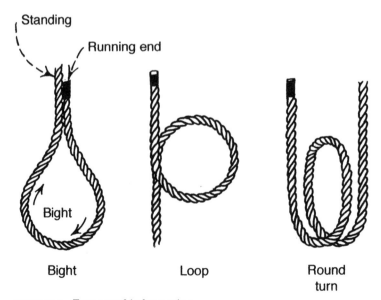

Figure 11.4   Terms used in knot tying.

TABLE 11.2 Polypropylene Rope Sling, Rated Capacity in Pounds* (Safety Factor-6)

Rope diameter (nominal), in.	Nominal weight per 100 ft. lb.	Minimum breaking strength, lb.	Eye and eye sling						Endless sling					
					Basket hitch						Basket hitch			
					Angle of rope to horizontal						Angle of rope to horizontal			
					90°	60°	45°	30°			90°	60°	45°	30°
					Angle of rope to vertical						Angle of rope to vertical			
			Vertical hitch	Choker hitch	0°	30°	45°	60°	Vertical hitch	Choker hitch	0°	30°	45°	60°
1/2	4.7	3,990	650	350	1,300	1,200	950	650	1,200	600	2,400	2,100	1,700	1,200
9/16	6.1	4,845	800	400	1,600	1,400	1,000	800	1,500	750	2,900	2,500	2,100	1,500
5/8	7.5	5,890	1,000	500	2,000	1,700	1,400	1,000	1,800	900	3,500	3,100	2,500	1,800
3/4	10.7	8,075	1,300	700	2,700	2,300	1,900	1,300	2,400	1,200	4,900	4,200	3,400	2,400
13/16	12.7	9,405	1,600	800	3,100	2,700	2,200	1,600	2,800	1,400	5,600	4,900	4,000	2,800
7/8	15.0	10,925	1,800	900	3,600	3,200	2,600	1,800	3,300	1,600	6,600	5,700	4,600	3,300
1	18.0	13,300	2,200	1,100	4,400	3,800	3,100	2,200	4,000	2,000	8,000	6,900	5,600	4,000
1 1/16	20.4	15,200	2,500	1,300	5,100	4,400	3,600	2,500	4,600	2,300	9,100	7,900	6,500	4,600
1 1/8	23.7	17,385	2,900	1,500	5,800	5,000	4,100	2,900	5,200	2,600	10,500	9,000	7,400	5,200
1 1/4	27.0	19,950	3,300	1,700	6,700	5,800	4,700	3,300	6,600	3,000	12,000	10,500	8,500	6,000
1 5/16	30.5	22,325	3,700	1,900	7,400	6,400	5,300	3,700	6,700	3,400	13,500	11,500	9,500	6,700
1 1/2	38.5	28,215	4,700	2,400	9,400	8,100	6,700	4,700	8,500	4,200	17,000	14,500	12,000	8,500
1 5/8	47.5	34,200	5,700	2,900	11,500	9,900	8,100	5,700	10,500	5,100	20,500	18,000	14,500	10,500
1 3/4	57.0	40,850	6,800	3,400	13,500	12,000	9,600	6,800	12,500	6,100	24,500	21,000	17,500	12,500
2	69.0	49,400	8,200	4,100	16,500	14,500	11,500	8,200	15,000	7,400	29,500	25,500	21,000	15,000
2 1/8	80.0	57,950	9,700	4,800	19,500	16,500	13,500	9,700	17,500	8,700	35,000	30,100	24,500	17,500
2 1/4	92.0	65,550	11,000	5,500	22,000	19,000	15,500	11,000	19,500	9,900	39,500	34,000	28,000	19,500
2 1/2	107.0	76,000	12,500	6,300	25,500	22,000	18,000	12,500	23,000	11,500	45,500	39,500	32,500	23,000
2 5/8	120.0	85,500	14,500	7,100	28,500	24,500	20,000	14,500	25,500	13,000	51,500	44,500	36,500	25,500

*See Fig. 11.3 for sling configuration descriptions.
Courtesy: Occupational Safety and Health Administration, *Rules and Regulations #1910.184 Slings,* 1976.

TABLE 11.3 Polyester Rope Sling, Rated Capacity in Pounds* (Safety Factor-9)

Rope diameter (nominal), in.	Nominal weight per 100 ft. lb.	Minimum breaking strength, lb.	Eye and eye sling						Endless sling						
					Basket hitch						Basket hitch			
					Angle of rope to horizontal						Angle of rope to horizontal			
			Vertical hitch	Choker hitch	90°	60°	45°	30°	Vertical hitch	Choker hitch	90°	60°	45°	30°
					Angle of rope to vertical						Angle of rope to vertical			
					0°	30°	45°	60°			0°	30°	45°	60°
1/2	8.0	6,080	700	350	1,400	1,200	950	700	1,200	600	2,400	2,100	1,700	1,200
9/16	10.2	7,600	850	400	1,700	1,500	1,200	850	1,500	750	3,000	2,600	2,200	1,500
5/8	13.0	9,500	1,100	550	2,100	1,800	1,500	1,100	1,900	950	3,800	3,300	2,700	1,900
3/4	17.5	11,875	1,300	650	2,600	2,300	1,900	1,300	2,400	1,200	4,800	4,100	3,400	2,400
13/16	21.0	14,725	1,600	800	3,300	2,800	2,300	1,500	2,900	1,500	5,900	5,100	4,200	2,900
7/8	25.0	17,100	1,900	950	3,800	3,300	2,700	1,900	3,400	1,700	6,800	5,900	4,800	3,400
1	30.5	20,900	2,300	1,200	4,600	4,000	3,300	2,300	4,200	2,100	8,400	7,200	5,900	4,200
1 1/16	34.5	24,225	2,700	1,300	5,400	4,700	3,800	2,700	4,800	2,400	9,700	8,400	6,900	4,800
1 1/8	40.0	28,025	3,100	1,600	6,200	5,400	4,400	3,100	5,600	2,800	11,000	9,700	7,900	5,600
1 1/4	46.3	31,540	3,500	1,800	7,000	6,100	5,000	3,500	6,300	3,200	12,500	11,000	8,900	6,300
1 5/16	52.5	35,625	4,000	2,000	7,900	6,900	5,600	4,000	7,100	3,600	14,500	12,500	10,000	7,100
1 1/2	66.8	44,460	4,900	2,500	9,900	8,600	7,000	4,900	8,900	4,400	18,000	15,500	12,500	8,900
1 5/8	82.0	54,150	6,000	3,000	12,000	10,400	8,500	6,000	11,000	5,400	21,500	19,000	15,500	11,000
1 3/4	98.0	64,410	7,200	3,600	14,500	12,500	10,000	7,200	13,000	6,400	26,000	22,500	18,000	13,000
2	118.0	76,000	8,400	4,200	17,000	14,500	12,000	8,400	15,000	7,600	30,500	26,500	21,500	15,000
2 1/8	135.0	87,400	9,700	4,900	19,500	17,000	13,500	9,700	17,500	8,700	35,000	30,500	24,500	17,500
2 1/4	157.0	101,650	11,500	5,700	22,500	19,500	16,000	11,500	20,500	10,000	40,500	35,000	29,000	20,500
2 1/2	181.0	115,900	13,000	6,400	26,000	22,500	18,000	13,000	23,000	11,500	46,500	40,000	33,000	23,000
2 5/8	205.0	130,150	14,500	7,200	29,000	25,000	20,500	14,500	26,000	13,000	52,000	45,000	37,000	26,000

*See Fig. 11.3 for sling configuration descriptions.
Courtesy: Occupational Safety and Health Administration. *Rules and Regulations #1910.184 Slings,* 1976.

TABLE 11.4 Nylon Rope Sling, Rated Capacity in Pounds* (Safety Factor-9)

Rope diameter (nominal), in.	Nominal weight per 100 ft. lb.	Minimum breaking strength, lb.	Eye and eye sling Vertical hitch	Eye and eye sling Choker hitch	Eye Basket 90° (vert 0°)	Eye Basket 60° (vert 30°)	Eye Basket 45° (vert 45°)	Eye Basket 30° (vert 60°)	Endless Vertical hitch	Endless Choker hitch	Endless Basket 90° (vert 0°)	Endless Basket 60° (vert 30°)	Endless Basket 45° (vert 45°)	Endless Basket 30° (vert 60°)
1/2	8.0	6,080	700	350	1,400	1,200	950	700	1,200	600	2,400	2,100	1,700	1,200
9/16	10.2	7,600	850	400	1,700	1,500	1,200	850	1,500	750	3,000	2,600	2,200	1,500
5/8	13.0	9,500	1,100	550	2,100	1,800	1,500	1,100	1,900	950	3,800	3,300	2,700	1,900
3/4	17.5	11,875	1,300	750	2,600	2,300	1,900	1,300	2,400	1,200	4,800	4,100	3,400	2,400
13/16	21.0	14,725	1,600	900	3,300	2,800	2,300	1,500	2,900	1,500	5,900	5,100	4,200	2,900
7/8	25.0	17,100	1,900	1,100	3,800	3,300	2,700	1,900	3,400	1,700	6,800	5,900	4,800	3,400
1	30.5	20,900	2,300	1,300	4,600	4,000	3,300	2,300	4,200	2,100	8,400	7,200	5,900	4,200
1 1/16	34.5	24,225	2,700	1,500	5,400	4,700	3,800	2,700	4,800	2,400	9,700	8,400	6,900	4,800
1 1/8	40.0	28,025	3,100	1,700	6,200	5,400	4,400	3,100	5,600	2,800	11,000	9,700	7,900	5,600
1 1/4	46.3	31,540	3,500	2,000	7,000	6,100	5,000	3,500	6,300	3,200	12,500	11,000	8,900	6,300
1 5/16	52.5	35,625	4,000	2,300	7,900	6,900	5,600	4,000	7,100	3,600	14,500	12,500	10,000	7,100
1 1/2	66.8	44,460	4,900	2,800	9,900	8,600	7,000	4,900	8,900	4,400	18,000	15,500	12,500	8,900
1 5/8	82.0	54,150	6,000	3,400	12,000	10,400	8,500	6,000	11,000	5,400	21,500	19,000	15,500	11,000
1 3/4	98.0	64,410	7,200	4,100	14,500	12,500	10,000	7,200	13,000	6,400	26,000	22,500	18,000	13,000
2	118.0	76,000	8,400	4,900	17,000	14,500	12,000	8,400	15,000	7,600	30,500	26,500	21,500	15,000
2 1/8	135.0	87,400	9,700	5,600	19,500	17,000	13,500	9,700	17,500	8,700	35,000	30,500	24,500	17,500
2 1/4	157.0	101,650	11,500	6,600	22,500	19,500	16,000	11,500	20,500	10,000	40,500	35,000	29,000	20,500
2 1/2	181.0	115,900	13,000	7,400	26,000	22,500	18,000	13,000	23,000	11,500	46,500	40,000	33,000	23,000
2 5/8	205.0	130,150	14,500	8,000	29,000	25,000	20,500	14,500	26,000	13,000	52,000	45,000	37,000	26,000

*See Fig. 11.3 for sling configuration descriptions.
Courtesy: Occupational Safety and Health Administration. *Rules and Regulations #1910.184 Slings,* 1976.

TABLE 11.5 Manila Rope Sling, Rated Capacity in Pounds* (Safety Factor-5)

Rope diameter (nominal), in.	Nominal weight per 100 ft. lb.	Minimum breaking strength, lb.	Eye and eye sling Vertical hitch	Eye and eye sling Choker hitch	Eye and eye sling Basket hitch 90° (0° vert)	Basket 60° (30° vert)	Basket 45° (45° vert)	Basket 30° (60° vert)	Endless sling Vertical hitch	Endless sling Choker hitch	Endless sling Basket hitch 90° (0° vert)	Basket 60° (30° vert)	Basket 45° (45° vert)	Basket 30° (60° vert)
1/2	7.5	2,650	550	250	1,100	900	750	550	950	500	1,900	1,700	1,400	950
9/16	10.4	3,450	700	350	1,400	1,200	1,000	700	1,200	600	2,500	2,200	1,800	1,200
5/8	13.3	4,400	900	450	1,800	1,500	1,200	900	1,600	800	3,200	2,700	2,200	1,600
3/4	16.7	5,400	1,100	550	2,200	1,900	1,500	1,100	2,000	950	3,900	3,400	2,800	2,000
13/16	19.5	6,500	1,300	630	2,600	2,300	1,800	1,300	2,300	1,200	4,700	4,100	3,300	2,300
7/8	22.5	7,700	1,500	750	3,100	2,700	2,200	1,500	2,800	1,400	5,600	4,800	3,900	2,800
1	27.0	9,000	1,800	900	3,600	3,100	2,600	1,800	3,200	1,600	6,500	5,600	4,600	3,200
1 1/16	31.3	10,500	2,100	1,100	4,200	3,600	3,000	2,100	3,800	1,900	7,600	6,600	5,400	3,800
1 1/8	36.0	12,000	2,400	1,200	4,800	4,200	3,400	2,400	4,300	2,200	8,600	7,500	6,100	4,300
1 1/4	41.7	13,500	2,700	1,400	5,400	4,700	3,800	2,700	4,900	2,400	9,700	8,400	6,900	4,900
1 5/16	47.9	15,000	3,000	1,500	6,000	5,200	4,300	3,000	5,400	2,700	11,000	9,400	7,700	5,400
1 1/2	59.9	18,500	3,700	1,850	7,400	6,400	5,200	3,700	6,700	3,300	13,500	11,500	9,400	6,700
1 5/8	74.6	22,500	4,500	2,300	9,000	7,800	6,400	4,500	8,100	4,100	16,000	14,000	11,500	8,000
1 3/4	89.3	26,500	5,300	2,700	10,500	9,200	7,500	5,300	9,500	4,800	19,000	16,500	13,500	9,500
2	107.5	31,000	6,200	3,100	12,500	10,500	8,800	6,200	11,000	5,600	22,500	19,500	16,000	11,000
2 1/8	125.0	36,000	7,200	3,600	14,500	12,500	10,000	7,200	13,000	6,500	26,000	22,500	18,500	13,000
2 1/4	146.0	41,000	8,200	4,100	16,500	14,000	11,500	8,200	15,000	7,400	29,500	25,500	21,000	15,000
2 1/2	166.7	46,500	9,300	4,700	18,500	16,000	13,000	9,300	16,500	8,400	33,500	29,000	23,000	16,500
2 5/8	190.8	52,000	10,500	5,200	21,000	18,000	14,500	10,500	18,500	9,500	37,500	32,500	26,500	18,500

*See Fig. 11.3 for sling configuration descriptions.
Courtesy: Occupational Safety and Health Administration. *Rules and Regulations #1910.184 Slings*, 1976.

Figure 11.5 The *overhand knot* is the simplest knot made and forms a part of many other knots. This knot is often tied at the end of a rope to prevent the strands from unraveling or as a stop knot to prevent rope from slipping through a block.

Figure 11.6 A *half hitch* is used to throw around the end of an object to guide it or keep it erect while hoisting. A half hitch is ordinarily used with another knot or hitch. A half hitch or two thrown around the standing part of a line after tying a clove hitch makes a very secure knot.

Figure 11.7 The *two half-hitches* knot is used in attaching a rope for anchoring or snubbing. It is easily and quickly made and easily untied.

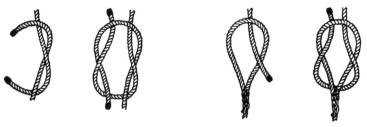

Figure 11.8 The *square knot* is used to tie two ropes together of approximately the same size. It will not slip and can usually be untied even after a heavy strain has been put on it. Linemen use the square knot to bind light leads, lash poles together on changeovers, on slings to raise transformers, and for attaching blocks to poles and crossarms.

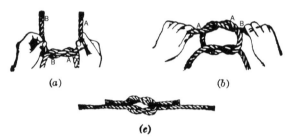

(a)   (b)

(c)

Figure 11.9 The method of making a *square knot*. (a) Passing left end A over right end B and under. (b) Passing right end B over left end A and under. (c) The completed knot drawn up. (*Courtesy of Plymouth Cordage Co.*)

Figure 11.10 When making a square knot, care must be taken that the standing and running parts of each rope pass through the loop of the other in the same direction, i.e., from above downward, or vice versa; otherwise a *granny knot* is made which will not hold.

Figure 11.11 In tying the square knot, the standing part of both ropes must cross, as otherwise a useless knot (known as the *thief knot*) is formed.

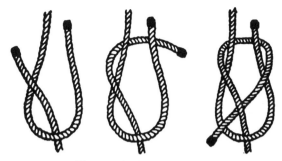

Figure 11.12   The *single sheet bend* is used in joining ropes, especially those of unequal size. It is more secure than the square knot but is more difficult to untie. It is made by forming a loop in one end of a rope; the end of the other rope is passed up through the loop and underneath the end and standing part, then down through the loop thus formed.

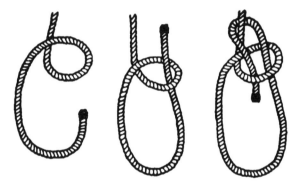

Figure 11.13   The *bowline* is used to place a loop in the end of a line. It will not slip or pull tight. Linemen use the bowline to attach come-alongs (wire grips) to rope, to attach tail lines to hook ladders, and (as a loose knot) to throw on conductors to hold them in the clear while working on poles.

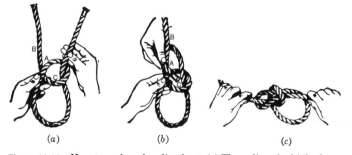

(a)   (b)   (c)

Figure 11.14   How to make a *bowline knot*. (a) Threading the bight from below. (b) Leading around standing part and back through bight C. (c) The completed bowline. (*Courtesy of Plymouth Cordage Co.*)

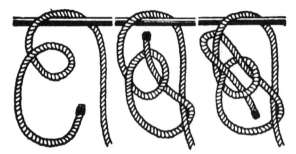

Figure 11.15 A *running bowline knot* is used when a hand line or bull rope is to be tied around an object at a point that cannot be safely reached, such as the end of a limb.

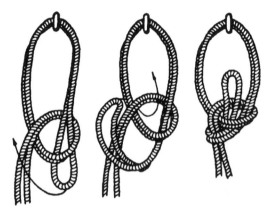

Figure 11.16 A *double bowline knot* is used to form a loop in the middle of a rope that will not slip when a strain is put upon it.

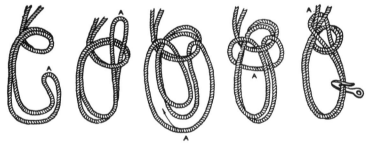

Figure 11.17 The *bowline on a bight* is used to place a loop in a line somewhere away from the end of the rope. It can be used to gain mechanical advantage in a rope guy by doubling back through the bowline on a bight much as a set of blocks. The bowline on a bight also makes a good seat for a man when he is suspended on a rope.

To tie this bowline, take the bight of the rope and proceed as with the simple bowline; only instead of tucking the end down through the bight of the knot, carry the bight over the whole and draw up, thus leaving it double in the knot and double in the standing part. The loop under the standing part is single.

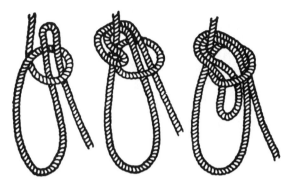

Figure 11.18 The *single intermediate bowline knot* is used in attaching rope to the hook of a block where the end of the rope is not readily available.

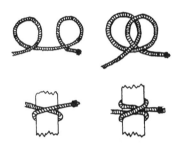

Figure 11.19 The *clove hitch* is used to attach a rope to an object such as a crossarm or pole where a knot that will not slip along the object is desired. Linemen use the clove hitch for side lines, temporary guys, and hoisting steel.

To make this hitch, pass the end of the rope around the spar or timber, then over itself, over and around the spar, and pass the end under itself and between the rope and spar as shown.

Figure 11.20 The *timber hitch* is used to attach a rope to a pole when the pole is to be towed by hand along the ground in places where it would be impossible to use a truck or its winch line to spot it. The timber hitch is sometimes used to send crossarms aloft. This hitch forms a secure temporary fastening which may be easily undone. It is similar to the half hitch but is more secure. Instead of the end being passed under the standing part of the rope once, it is wound around the standing part three or four times, as shown in the figure.

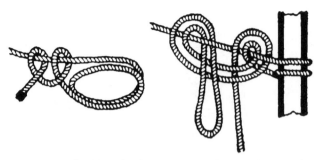

Figure 11.21 *Snubbing hitch knots* are used to attach a rope for anchoring or snubbing purposes.

Figure 11.22 The *taut-rope hitch knot* is used to attach one rope to another for snubbing a load.

Figure 11.23 The *rolling bend* is often used to attach the rope to wires that are too large for the wire-pulling grips. It is also used for skidding poles or timber.

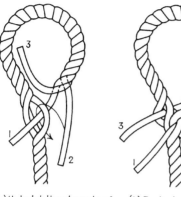 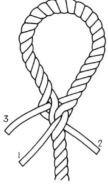 

(a) Untwist the strands of the rope for a length of about 12 in. Throw a bight the size of the required eye into the rope. Tuck strand 1 as shown and cross strand 2 behind it

(b) Tuck strand 3 behind as shown and pull all strands tight

(c) Tuck strand 1 by passing it over the adjacent strand and under the next one

(e) Tuck strand 2. Pull all strands tight. Continue until three tucks for natural fibers and four tucks for synthetic fiber ropes have been made in each strand. To taper the splice, split the strands and remove one-third of each. Make one tuck with the remaining two-thirds of each strand. Remove half the remainder of each strand and make a final tuck. Roll the splice between your foot and the floor to smooth out. Cut off surplus ends flush with outside strands

(d) Similarly tuck strand 3

Figure 11.24   Eye splice, fiber rope.

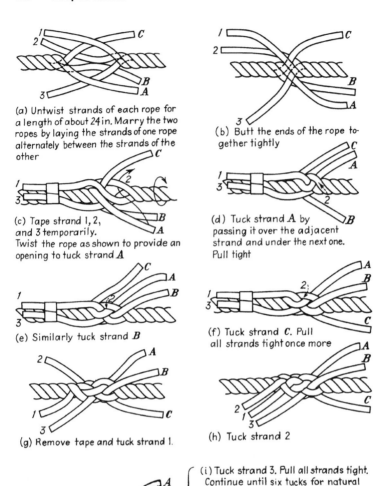

(a) Untwist strands of each rope for a length of about 24 in. Marry the two ropes by laying the strands of one rope alternately between the strands of the other

(b) Butt the ends of the rope together tightly

(c) Tape strand 1, 2, and 3 temporarily. Twist the rope as shown to provide an opening to tuck strand A

(d) Tuck strand A by passing it over the adjacent strand and under the next one. Pull tight

(e) Similarly tuck strand B

(f) Tuck strand C. Pull all strands tight once more

(g) Remove tape and tuck strand 1.

(h) Tuck strand 2

(i) Tuck strand 3. Pull all strands tight. Continue until six tucks for natural fiber and eight tucks for synthetic fiber ropes have been made in each strand. To taper the splice, split the strands and remove one-third of each. Make one tuck with the remaining two-thirds of each strand. Remove half of the remainder of each strand and make a final tuck. Roll the splice between your foot and the floor to smooth out. Cut off surplus ends flush with outside strands

Figure 11.25   Short splice, fiber rope.

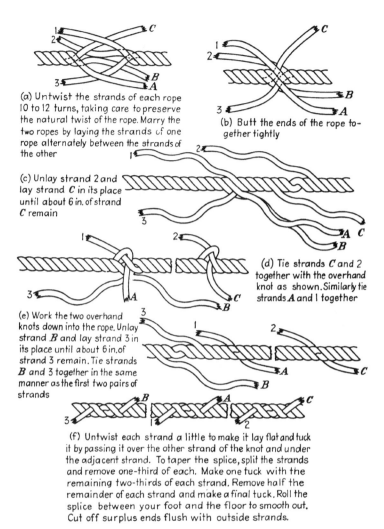

(a) Untwist the strands of each rope 10 to 12 turns, taking care to preserve the natural twist of the rope. Marry the two ropes by laying the strands of one rope alternately between the strands of the other

(b) Butt the ends of the rope together tightly

(c) Unlay strand 2 and lay strand C in its place until about 6 in. of strand C remain

(d) Tie strands C and 2 together with the overhand knot as shown. Similarly tie strands A and 1 together

(e) Work the two overhand knots down into the rope. Unlay strand B and lay strand 3 in its place until about 6 in. of strand 3 remain. Tie strands B and 3 together in the same manner as the first two pairs of strands

(f) Untwist each strand a little to make it lay flat and tuck it by passing it over the other strand of the knot and under the adjacent strand. To taper the splice, split the strands and remove one-third of each. Make one tuck with the remaining two-thirds of each strand. Remove half the remainder of each strand and make a final tuck. Roll the splice between your foot and the floor to smooth out. Cut off surplus ends flush with outside strands.

Figure 11.26 Long splice, fiber rope. The long splice is used to permanently join two ropes which must pass through a close-fitting pulley.

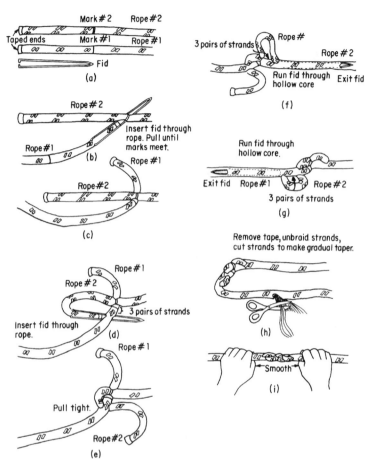

Figure 11.27 In-line splice for 12-strand braided synthetic fiber rope. (*Courtesy of Yale Cordage, Inc.*)

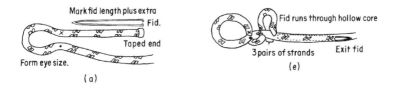

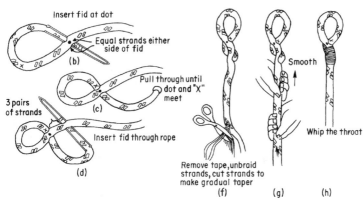

Figure 11.28 An eye splice for 12-strand braided synthetic fiber rope. (*Courtesy Yale Cordage, Inc.*)

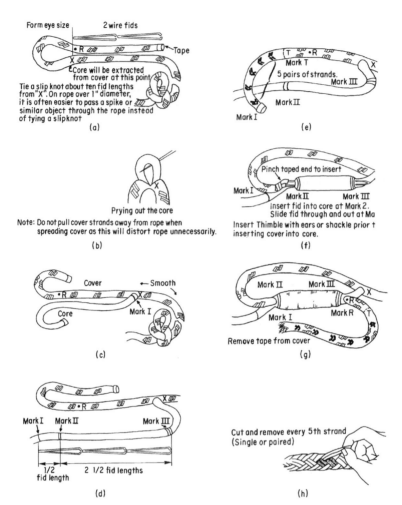

Figure 11.29   An eye splice for double-braided synthetic fiber rope. (*Courtesy of Yale Cordage, Inc.*)

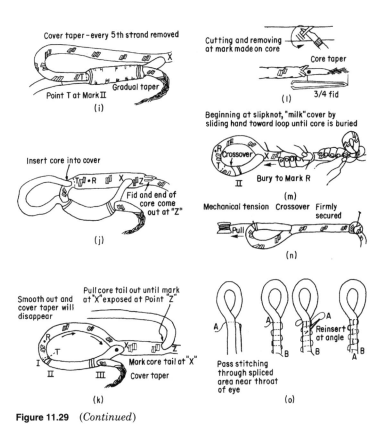

Figure 11.29   (*Continued*)

TABLE 11.6   Percentage Strength of Spliced or Knotted Rope

Type of splice or knot	Percentage strength
Straight rope	100
Eye splice	90
Short splice	80
Timber hitch or half hitch	65
Bowline or clove hitch	60
Square knot or sheet bend	50
Overhand knot (half or square knot)	45

LOAD LIFTED	DIAGRAM OF·RIGGING
2 times safe load on rope (approx.) **	Single ⬤ block 2 parts ⟵ Single ⬤ block 
3 times safe load on rope (approx.) **	Double ⬤ block 3 parts ⟵ Single ⬤ block 
4 times safe load on rope (approx.) **	Double ⬤ block 4 parts ⟵ Double ⬤ block 
5 times safe load on rope (approx.) **	Triple ⬤ block 5 parts Double ⬤ block 
6 times safe load on rope (approx.) **	Triple ⬤ block 6 parts Triple ⬤ block 
7 times safe load on rope (approx.) **	Quadruple ⬤ block 7 parts Triple ⬤ block 

Figure 11.30 Lifting capacity of block and tackle.
**Less 20 percent approximately for friction.

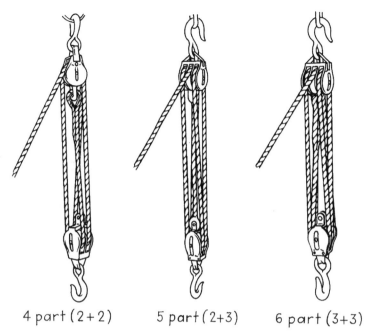

4 part (2+2)   5 part (2+3)   6 part (3+3)

Figure 11.31 Four-, five-, and six-part blocks and tackle.

NOTES

Grounding

Codes

The *National Electrical Safety Code* (ANSI-C2 Standard) specifies the grounding methods for electrical supply and communications facilities.

The *National Electrical Code*® (NFPA-70 Standard) specifies the grounding methods for all facilities that are within or on public or private buildings or other structures.

Definitions

Ground   The means by which currents are conducted to remote earth. Proper grounding for electrical systems is very important for the safety of the linemen, cablemen, groundmen, and the public. Correct operation of protective devices is dependent on the grounding installation.

Ground resistance   The electrical resistance of the earth is largely determined by the chemical ingredients of the soil and the amount of moisture present (Figs. 12.1 through 12.3). Measurements of ground resistances completed by the Bureau of Standards are summarized in Table 12.1.

Ground Rods

If the ground rod is driven to its full length and the ground resistance is above 25 ohms, resort to one of the following procedures:

- Extend the length of the rod.
- Drive additional rods.
- Treat the soil surrounding the rod with chemicals.

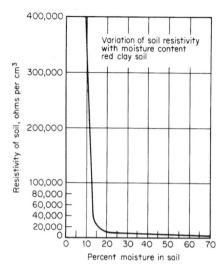

Figure 12.1 Variation of soil resistivity with moisture content. (*Courtesy of Copperweld Steel Co.*)

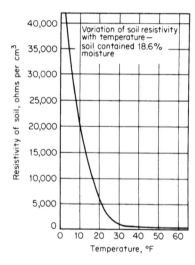

Figure 12.2 Variation of soil resistivity with temperature. (*Courtesy of Copperweld Steel Co.*)

Effects of multiple rods tied together to lower resistance can be seen in Figs. 12.4 and 12.5:

- When two or more driven rods are well spaced from each other and connected together, they provide parallel paths to the earth.

- They become, in effect, resistances in parallel or multiple and tend to follow the law of metallic parallel resistances.

- The rods are connected together with wire no smaller than that used to connect to the top of the pole.

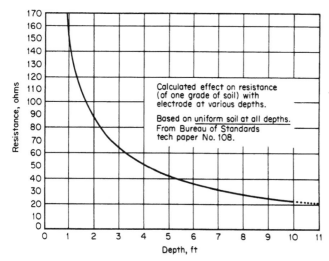

Figure 12.3   Chart showing the relationship between depth and resistance for a soil with a uniform moisture content at all depths. In the usual field condition, deeper soils have a higher moisture content and the advantage of depth is more pronounced. (*Courtesy of Copperweld Steel Co.*)

TABLE 12.1   **Resistance of Different Types of Soil**

	Resistance, ohms		
Soil	Average	Minimum	Maximum
Fills and ground containing more or less refuse such as ashes, cinders, and brine waste	14	3.5	41
Clay, shale, adobe, gumbo, loam, and slightly sandy loam with no stones or gravel	24	2.0	98
Clay, adobe, gumbo, and loam mixed with varying proportions of sand, gravel, and stones	93	6.0	800
Sand, stones, or gravel with little or no clay or loam	554	35	2700

Courtesy of Bureau of Standards Technologic Paper 108.

- Multiple rods are commonly used for arrester grounds and for station and substation grounds.
- In addition to lowering the resistance, multiple rods provide higher current-carrying capacity and can thus handle larger fault currents.

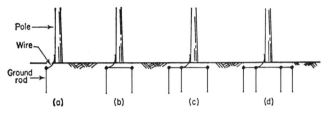

Figure 12.4 One ground rod (a) and two, three, and four ground rods are shown at poles (b), (c), and (d), respectively. Rods at each pole are connected together electrically.

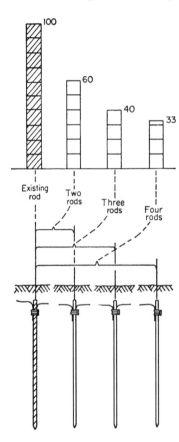

Figure 12.5 Installations of multiple ground rods. The graphs show the approximate extent to which ground resistance is reduced by the use of two, three, or four rods connected in multiple. If the rods were widely separated, two rods would have 50 percent reduction of the resistance of one rod, but because they are usually driven only 6 ft or more from each other, the resistance of two rods is about 60 percent of that of one rod. Three rods, instead of having 33 percent resistance of one rod, have on an average about 40 percent. Four rods, instead of having 25 percent resistance of one rod, have about 33 percent.

Transmission-Line Grounds

Figures 12.6 through 12.9 show various types of transmission-line grounds.

■ When lightning strikes an overhead ground or static wire on a transmission line, the lightning current is conducted to ground through the metal tower or the ground wire installed along the pole.

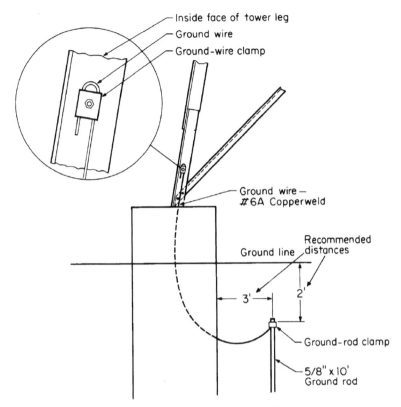

Figure 12.6   A ground wire connects the ground rod to the metal tower.

- The top of the structure is raised in potential to a value determined by the magnitude of the lightning current and the surge impedance of the ground connection.

- If the impulse resistance of the ground connection is high, this potential can be many thousands of volts. If the potential exceeds the insulation level of the equipment, flashover will result, causing a power arc that will initiate the operation of protective relays and removal of the line from service.

- If the transmission structure is well grounded and proper coordination exists between the ground resistance and the conductor insulation, flashover can usually be avoided.

- The transmission-line grounds are typically installed at each structure to obtain a low ground resistance.

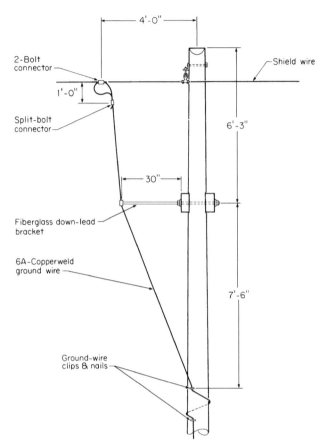

Figure 12.7 A shield-wire connection to the ground wire on the wood pole for the transmission line.

Grounding Coils and Grounding Plates (Figs. 12.10 through 12.12)

- A grounding coil is a spiral coil of bare copper wire placed at the bottom of a pole.

- The wire of the coil continues up the pole as the ground-wire lead.

- Sometimes the coil is wound around the lower end of the pole's side a few times as a helix to increase the amount of wire surface in contact with the earth (Fig. 12.11).

- Grounding coils should have enough turns to make good contact with the earth.

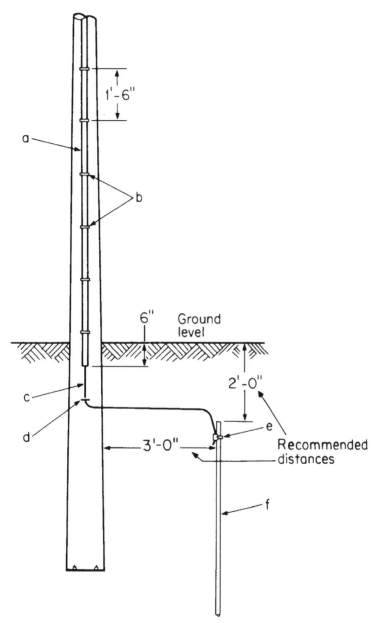

Figure 12.8   A ground-wire connection to the ground rod for a wood-pole transmission-line structure. (a) Ground-wire molding; (b) staples, molding; (c) 6A Copperweld wire; (d) staples, fence; (e) ground-rod clamp; and (f) 5/8-in × 10-ft ground rod.

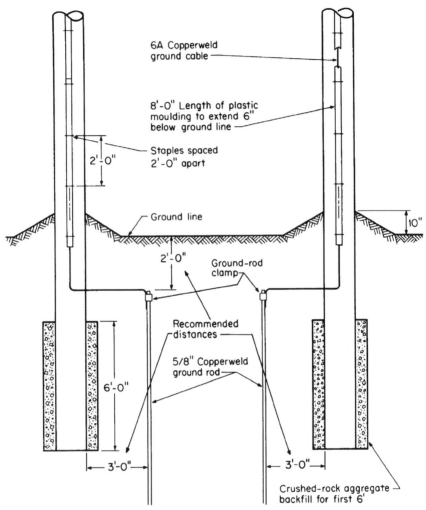

Figure 12.9 Ground-wire connections to ground rods for a wood-pole H-frame transmission-line structure.

- A coil having seven turns and 13 ft or more of wire would provide satisfactory ground contact (Fig. 12.12).
- Number 6 or larger AWG soft-drawn or annealed copper wire should be used.
- To prevent the coil from acting as a choke coil, the ground lead from the innermost turn is stapled to all the other turns.
- An alternative to the grounding coil is the grounding plate (Fig. 12.10).

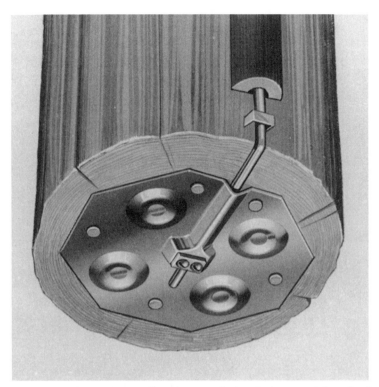

Figure 12.10   A copper grounding plate attached to the bottom of the pole of a transmission or distribution line before the pole is set. The grounding plate has moisture-retaining cups to maintain a good ground. (*Courtesy of Homac Manufacturing Co.*)

Figure 12.11   A grounding coil in position on the bottom of the pole. The entire weight of the pole helps to maintain contact with the earth below the pole. The spiral coil is stapled to the bottom of the pole. Wire from the inner spiral is stapled to all turns as it crosses them to shunt out the turns.

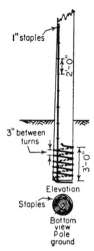

Figure 12.12 Views of the side and bottom grounding coils. The elevation shows six twin helix with 3 in between turns. The bottom view shows a spiral pancake coil of seven turns. (*Courtesy of Homac Manufacturing Co.*)

Counterpoise

- If the soil has a high resistance, a grounding system called a counterpoise may be necessary (Fig. 12.13).

- The counterpoise for an overhead transmission line consists of a special grounding terminal that reduces the surge impedance of the ground connection and increases the coupling between the ground wire and the conductors.

- Counterpoises are normally installed for transmission-line structures located in areas with sandy soil or rock close to the surface.

- The types of counterpoises used are the continuous (parallel) type and the radial (crowfoot) type.

- The continuous (parallel) counterpoise consists of one or more conductors buried under the transmission line for its entire length or under sections with high-resistance soils.

- The counterpoise wires are connected to the overhead ground (static wire) at all supporting structures. The radial-type counterpoise consists of a number of wires extending radially from the tower legs.

- The number and length of the wires will depend on the tower location and the soil conditions.

- The continuous counterpoise wires are usually installed with a cable plow at a depth of 18 inches or more.

- The wires should be deep enough so that they will not be disturbed by cultivation of the land.

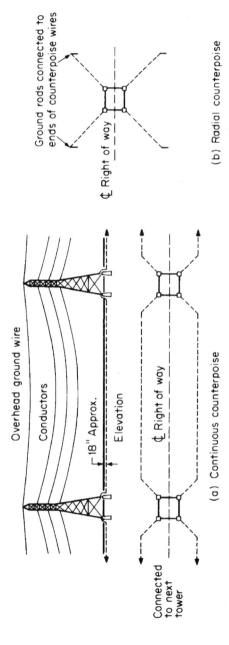

Overhead ground wire

Conductors

18" Approx.

Elevation

Connected to next tower

¢ Right of way

(a) Continuous counterpoise

Ground rods connected to ends of counterpoise wires

¢ Right of way

(b) Radial counterpoise

Figure 12.13 Two general arrangements of counterpoise installations. (*Courtesy of Copperweld Steel Co.*)

Distribution-Circuit Grounds
(Figs. 12.14 through 12.17)

- A multigrounded neutral conductor for a primary distribution line is always connected to the substation grounding system where the circuit originates and is connected to all grounds along the length of the circuit.

- The primary neutral conductor must be continuous over the entire length of the circuit and should be grounded at intervals not to exceed 1/4 mile.

- The primary neutral conductor can serve as a secondary neutral conductor.

- If separate primary and secondary neutral conductors are installed, the conductors should be connected together if the primary neutral conductor is effectively grounded.

- All equipment cases and tanks should be grounded.

- Lightning-arrester ground terminals should be connected to the common neutral, if available, and to a separate ground rod installed as close as is practical.

- Switch handles, down guys, and transformer secondaries must be properly grounded.

- The common conductor on the transformer secondary for a single-phase 120/240-volt service must be grounded.

- A three-phase delta- or open-delta-connected transformer secondary for 240- or 480-volt service would have the center tap of one of the transformer secondary terminals grounded.

- If the center tap of one of the transformer secondaries is not available, one-phase conductor is normally grounded.

- The neutral wire for three-phase, four-wire Y-connected secondaries for 208Y/120-volt or 480Y/277-volt service is grounded.

- If the primary circuit has an effectively grounded neutral conductor, the primary and secondary neutrals are connected together as well as to ground.

- Many times the primary distribution-line grounded neutral conductor is used for the secondary grounded neutral conductor and is called a common grounded neutral conductor.

- The neutral conductor of the service to the customer must be connected to ground on the customer's premises at the service-entrance equipment.

- A metal water pipe provides a good ground.

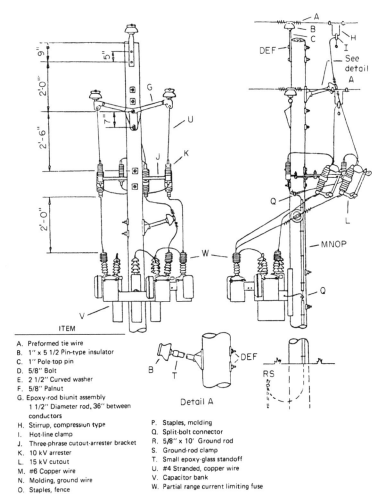

ITEM

A. Preformed tie wire
B. 1" x 5 1/2 Pin-type insulator
C. 1" Pole-top pin
D. 5/8" Bolt
E. 2 1/2" Curved washer
F. 5/8" Palnut
G. Epoxy-rod biunit assembly
   1 1/2" Diameter rod, 36" between
   conductors
H. Stirrup, compression type
I. Hot-line clamp
J. Three-phrase cutout-arrester bracket
K. 10 kV arrester
L. 15 kV cutout
M. #6 Copper wire
N. Molding, ground wire
O. Staples, fence

P. Staples, molding
Q. Split-bolt connector
R. 5/8" x 10' Ground rod
S. Ground-rod clamp
T. Small epoxy-glass standoff
U. #4 Stranded, copper wire
V. Capacitor bank
W. Partial range current limiting fuse

Figure 12.14 Capacitor bank installation for 13.8-kV circuit with grounding details illustrated.

■ Care must be taken to avoid plastic, cement-lined, or earthenware piping.

■ Driven grounds may be installed near the customer's service entrance to provide the required ground connection.

Installing Driven Grounds

■ A galvanized-steel rod, stainless-steel rod, or copperweld rod is driven into the ground beside the pole at a distance of 2 ft.

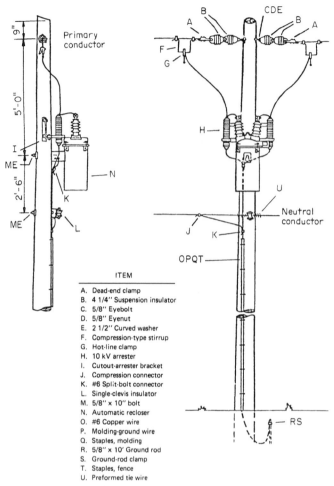

Primary conductor

Neutral conductor

ITEM

A. Dead-end clamp
B. 4 1/4" Suspension insulator
C. 5/8" Eyebolt
D. 5/8" Eyenut
E. 2 1/2" Curved washer
F. Compression-type stirrup
G. Hot-line clamp
H. 10 kV arrester
I. Cutout-arrester bracket
J. Compression connector
K. #6 Split-bolt connector
L. Single-clevis insulator
M. 5/8" x 10" bolt
N. Automatic recloser
O. #6 Copper wire
P. Molding-ground wire
Q. Staples, molding
R. 5/8" x 10' Ground rod
S. Ground-rod clamp
T. Staples, fence
U. Preformed tie wire

Figure 12.15 A single-phase 15-kV rural-distribution automatic-recloser installation with grounding details illustrated.

- The rod should be driven until the top end is 2 ft below the ground level.

- After the rod is driven in place, connection is made with the ground wire from the pole.

- The actual connection to the ground rod is usually made with a heavy bronze clamp.

- The clamp is placed over the ground-rod end and the ground-wire end.

- The size of the ground wire should not be less than number 6 AWG copper wire.

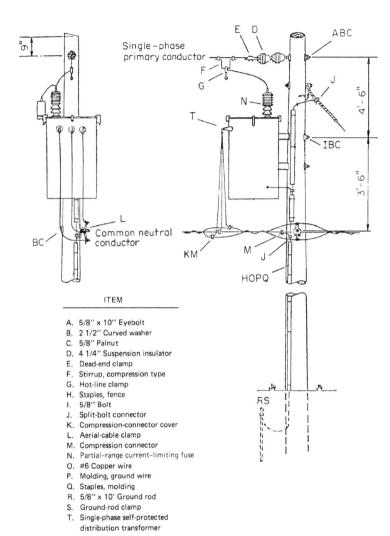

ITEM

A. 5/8" x 10" Eyebolt
B. 2 1/2" Curved washer
C. 5/8" Palnut
D. 4 1/4" Suspension insulator
E. Dead-end clamp
F. Stirrup, compression type
G. Hot-line clamp
H. Staples, fence
I. 5/8" Bolt
J. Split-bolt connector
K. Compression-connector cover
L. Aerial-cable clamp
M. Compression connector
N. Partial-range current-limiting fuse
O. #6 Copper wire
P. Molding, ground wire
Q. Staples, molding
R. 5/8" x 10' Ground rod
S. Ground-rod clamp
T. Single-phase self-protected
   distribution transformer

Figure 12.16   Distribution transformer installed on dead-end pole for 15-kV line with grounding details illustrated.

- Number 4 wire is generally used.

- When the setscrew in the clamp is tightened, the ground wire is squeezed against the ground wire.

- Several precautions are taken in making the actual connection to minimize the possibility of the ground wire being pulled out by frost action or packing of the ground above it.

- The ground wire may be brought up through the clamp from the bottom.

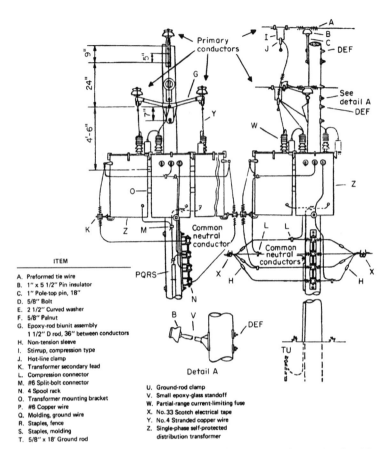

ITEM

A. Preformed tie wire
B. 1" x 5 1/2" Pin insulator
C. 1" Pole-top pin, 18"
D. 5/8" Bolt
E. 2 1/2" Curved washer
F. 5/8" Palnut
G. Epoxy-rod biunit assembly
   1 1/2" D rod, 36" between conductors
H. Non-tension sleeve
I. Stirrup, compression type
J. Hot-line clamp
K. Transformer secondary lead
L. Compression connector
M. #6 Split-bolt connector
N. 4 Spool rack
O. Transformer mounting bracket
P. #6 Copper wire
Q. Molding, ground wire
R. Staples, fence
S. Staples, molding
T. 5/8" x 18' Ground rod

U. Ground-rod clamp
V. Small epoxy-glass standoff
W. Partial-range current-limiting fuse
X. No. 33 Scotch electrical tape
Y. No. 4 Stranded copper wire
Z. Single-phase self-protected
   distribution transformer

Figure 12.17 Three-phase Y-Y 15-kV distribution transformer bank with quadruplex secondaries. The example depicts a common neutral conductor system.

- It is then bent over the clamp so that it cannot pull out.
- The wire is trained loosely so that there is plenty of slack between the ground-wire molding and the ground rod.

Ground-Wire Molding

- The molding protects the wire as well as the lineman.
- The molding should extend well down below the surface of the ground so that in case the ground rod bakes out, that is, fails to make contact with moist earth, there will not be any chance of a child or animal coming in contact with the ground wire and the ground at the same time.

■ When the installation is finally covered up, the ground wire should be completely out of sight.

■ Another reason for covering the ground wire is to protect persons in case of lightning-arrester trouble. If a lightning arrester should break down and allow power current to flow, the ground wire will become "hot" and remain so for an indefinite period of time if the resistance of the ground connection is high. Anyone touching the wire while it is energized could get an electric shock.

NOTES

13

Protective Grounds

Reference Standard

ASTM F855—Standard Specifications for Temporary Protective Grounds to Be Used on De-energized Electric Power Lines and Equipment

- Protection of the lineman is most important when a transmission or distribution line or a portion of a line is removed from service to be worked on using de-energized procedures.

- Using de-energized procedures, precautions are taken to ensure the line is de-energized before the work is started and remains de-energized until the work is completed.

- A line is grounded and short-circuited whenever it is to be worked on using de-energized procedures.

- De-energized lines are grounded for the protection of all the men working on the line.

- The protective grounds are installed from ground in a manner to short-circuit the conductors so that the linemen and everything in the working area will be at equal potential. This is the "man-shorted-out" concept, because the grounding, short-circuiting, and bonding leads will carry any current that may appear because of potential differences in the work area.

- When a line is to be worked on de-energized, the line must be grounded and short-circuited at the work location, even though the work location is within sight of the disconnecting means used to de-energize the line.

- When work is to be done at more than one location in a line section, the line section being worked on must be grounded and short-circuited at one location, and only the conductor(s) being worked on must be grounded at the work location.

- When a conductor is to be cut or opened, the conductor must be grounded on both sides of the open or grounded on one side of the open and bridged with a jumper cable across the point to be opened.

- A cluster block (Fig. 13.1) can be installed on a wood-pole structure to facilitate the ground connections and to help maintain the equal-potential area on the pole.

Figure 13.1 A grounding cluster installed on four-wire, three-phase line. The line conductor mounted on side of pole is the grounded neutral conductor. The grounding conductors create a low-impedance short circuit around the line-man's work area. (*Courtesy of A. B. Chance Co.*)

- Lines are grounded before the work is started and must remain so until all work is completed.

- The same precautions apply to new lines when construction has progressed to the point where they can be energized from any source.

- The *IEEE Guide for Protective Grounding of Power Lines* (IEEE Standard 1048) provides additional information about the installation of protective grounds.

The installation of protective grounds and short-circuiting leads at the work site protect against

- The hazards of static charges on the line
- Induced voltages
- Accidental energizing of the line

When a de-energized line and an energized line parallel each other:

- The de-energized line could pick up a static charge from the energized line because of the proximity of the lines.

- The de-energized line could have a voltage induced on it in the same manner as the secondary of a transformer.

- The amount of this static voltage "picked up" on the de-energized line depends on the length of the parallel, weather conditions, and many other variable factors.

- It could be hazardous, and precautions must be taken to protect against it by grounding the line at the location where the work is to be completed.

- If the de-energized line is grounded at a location remote from where the work is being done, this induced voltage will be present at the work location.

- Grounding the line at the work location will eliminate danger from induced voltages.

- Grounding at the work site will drain any static voltage to ground and protect the lineman from this potential hazard.

- Grounding and short-circuiting protect against the hazard of the line becoming energized from either accidental closing in of the line or accidental contact with an energized line that crosses or is adjacent to the de-energized line.

- If a circuit should be inadvertently energized, the grounds and short circuits on the line will cause the protective relays to initiate tripping of the circuit breaker at the source end of the energized line in a fraction of a second and de-energize the hot line.

■ During this short interval of time, the grounds and short circuits on the line being worked on will protect the linemen.

If it isn't grounded, it isn't dead!

Testing before grounding:

■ The lineman must test each line conductor to be grounded to ensure that it is de-energized (dead) before the protective grounds are installed.

■ The circuit can be tested with a high-voltage voltmeter or other special testing equipment that produces an audible sound.

■ The high-voltage voltmeter connected between ground and conductor energized with a static charge will give an indication of voltage and then will gradually indicate that voltage has reduced to zero, as the static charge is drained off to ground through the high-voltage voltmeter.

■ Each conductor of the line must be tested to ensure it is de-energized immediately prior to installing the grounding conductors.

Protective Ground Installation

After the testing is completed, the protective grounds should be installed in the following sequence:

1. Connect one end of the grounding bushing to an established ground. The grounded neutral conductor on a distribution line and the grounded static wire on a transmission line are the best sources of an established ground. A driven ground rod at the work site connected to the neutral conductor or static wire provides additional protection. If the work is being performed on a wood-pole structure, the installation of a cluster block on the pole below the work area connected to the grounding conductor, will ensure the establishment of an equal-potential zone on the pole for the lineman performing the work.

2. Connect the other end of the grounding conductor, using a hot-line tool, to the bottom conductor on vertical constructions or the closest conductor on horizontal constructions.

3. Install grounds or jumpers from a grounded conductor to the ungrounded conductors in sequence until all conductors are grounded and short-circuited together.

4. When a connection is made between a phase conductor and ground, the grounding lead, the connection to the earth, and the earth itself becomes a part of the electric circuit.

5. When the work is completed and the protective grounds are removed, always remove the grounds from the line conductors in reverse sequence, removing the connection to ground last (see Fig. 13.2).

Figures 13.3 through 13.5 show graphs of fault currents and fusing times for various types of conductors.

Figure 13.2 A lineman completing the installation of protective grounds using a hot-line tool. The grounding operation was completed by connecting the cluster block, mounted on the pole below the work area, to a driven ground rod. Then a grounding jumper was connected from the cluster block to one phase wire. Jumpering was completed in sequence from bottom to top, providing the lineman with a zone of protection as the protective grounds are installed. The grounded static wire at the top of the pole provides an effective ground source. (*Courtesy of A. B. Chance Co.*)

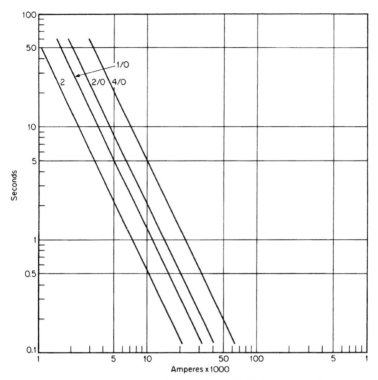

Figure 13.3 The suggested maximum allowable fault currents for copper grounding cables. The total current-on time, for minimum off time between consecutive current-on periods-reclosures. These values exceed IPCEA recommendations for cable installations by 1.91. (Based on tests of 10- to 60-cycle duration, 30°C ambient.) (*Courtesy of A. B. Chance Co.*)

Protective Ground Requirements

- A low-resistance path to the earth
- Clean connections
- Tight connections
- Connections made to proper points
- Adequate current-carrying capacity of grounding equipment

Sources of Ground

- A good available ground (earth connection) is the system neutral; however, it usually is not available on a transmission line.
- If a common neutral is at the location where the line is to be grounded, it must be connected to the grounding system on the structure so that all grounds in the immediate vicinity are connected together.

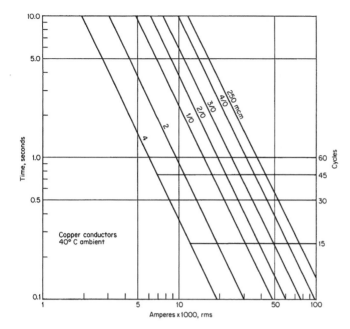

Figure 13.4   Fusing current time for copper conductors, 40°C ambient temperature. (*Courtesy of A. B. Chance Co.*)

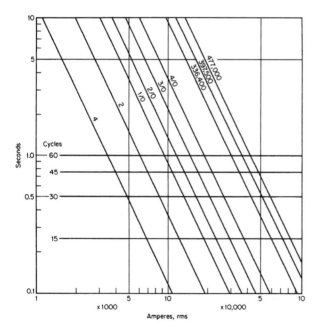

Figure 13.5   Fusing currents, aluminum conductor, 40°C ambient temperature. (*Courtesy of A. B. Chance Co.*)

- If the structure is metal, the neutral should be bonded to the steel—either with a permanent connection or a temporary jumper.

- If the structure is a wood pole that has a ground-wire connection to the static wire, the neutral should be permanently connected to the ground wire.

- If there is no permanent connection, a temporary jumper should be installed.

- If no ground wire is on the pole, the common neutral should be temporarily bonded to whatever temporary grounding connection is used to ground the line.

- Static wires are connected to the earth at many points along the line and provide a low-resistance path for grounding circuits.

- These are normally available for use as the ground-end connection of the grounding leads.

- If the line has a static wire, it is usually grounded at each pole.

- A ground wire runs down the wood pole(s) and connects to the ground rod.

- All metal poles and towers are adequately grounded when they are installed and serve as a good ground-connection point, provided the connection between the metal structure and the ground rod has not been cut or disconnected.

- If an exposed lead runs from the structure to the ground, it should be checked for continuity.

- If a neutral conductor or static wire grounded at the structure is not available, use any good metallic object, such as an anchor rod or a ground rod, which obviously extends several feet into the ground as a ground for the protective ground source.

- If no such ground is available, the lineman must provide a ground by driving a ground rod or screwing down a temporary anchor until it is firm and in contact with moist soil.

Figures 13.6 through 13.11 show various types of ground connections.

Underground-System Protective Grounds

- Underground transmission and distribution cable conductors are not readily accessible for grounding.

- Transmission cables originating and terminating in substations can normally be grounded with portable protective grounding equipment at the cable termination points in the same manner as described for overhead transmission circuits.

Figure 13.6   A running ground connection is designed to keep the conductor grounded as stringing operations are in process. A rope is used to secure the connector in place. (*Courtesy of Sherman & Reilly, Inc.*)

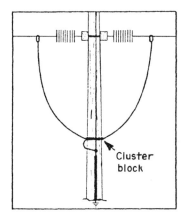 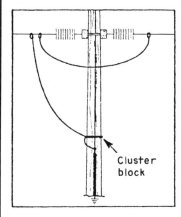

Cluster block

Cluster block

Figure 13.7   A method to ground and bridge conductors across an open point.

- If the cable termination points are not accessible, special equipment manufactured and provided for grounding the circuit must be used.
- Underground distribution cables that originate in substation switchgear to supply complete underground circuits must be tested and grounded with special equipment provided at the substation before work is

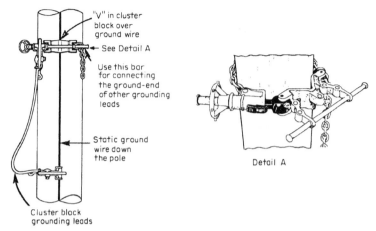

Figure 13.8   A metal cluster block equipped with 4-ft chain binder. The connection bar will accommodate four ground-lead clamps.

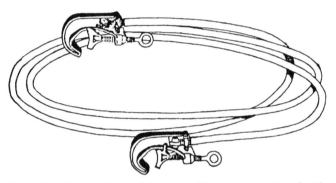

Figure 13.9   A grounding and short-circuiting jumper equipped with C-type clamps designed for operation with hot-line tool. The jumper cable is 1/0 copper conductor or larger with a large number of strands to provide flexibility.

performed on the cable conductors. Underground equipment installed in vaults may be equipped with oil switches that can be used to isolate and ground the cable conductors.

- Underground cable circuits originating at a riser pole in an overhead line can be grounded at the riser pole with portable protective grounding equipment.

- Underground cable circuits originating in pad-mounted switchgear can be grounded at the switchgear near the cable termination. The switches in the switchgear must be open to isolate the cable circuit. Opening the switch provides a visible break in the circuit.

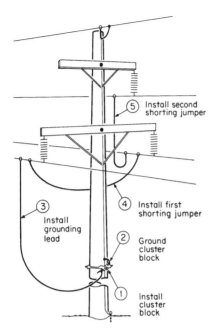

⑤ Install second
shorting jumper

③ Install
grounding
lead

④ Install first
shorting jumper

② Ground
cluster
block

① Install
cluster
block

Figure 13.10 Protective grounds installed on conductors of a single-circuit wood-pole sub-transmission line.

- A high-voltage voltmeter detecting tool is used by the lineman to be sure the circuit does not have a feedback from the remote end, proving that the cables are de-energized.

- A special grounding elbow and conductor is available above the cable terminations for the application of portable grounding devices in the switchgear.

- Underground-cable conductors terminating in pad-mounted transformers can be isolated, tested, and grounded at the transformer location.

- The lineman opens the pad-mounted transformer-compartment door and installs a feed-through device with a hot-line Grip-All clamp stick to prepare for the insulating and grounding operation.

- The cable to be worked on must be isolated to separate it from a source of energy at the remote end.

- The lineman or cableman identifies the cable to be worked on at the pad-mount transformer with the markings on the cable and maps of the circuit.

- When the proper cable has been identified, the elbow cable terminator can be separated from the bushing on the transformer with the Grip-All clamp stick.

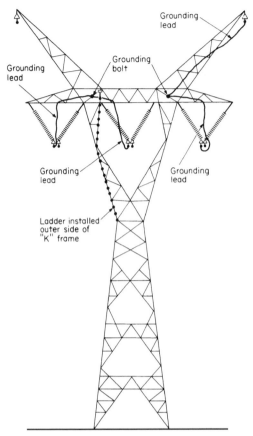

Figure 13.11   Protective grounds installed on conductors of a single-circuit transmission line constructed on steel towers.

- The lineman then installs the elbow on the bushing on the feed-through device with the hot stick.

- A high-voltage tester equipped with a special connection can be used to test the isolated cable through the second bushing on the feed-through device.

- If the cable is found to be de-energized, the lineman can connect a special grounding conductor to the ground wires in the pad-mounted transformer compartment and place it on the bushing of the feed-through device with a Grip-All clamp stick.

- The cable is now properly grounded so that the necessary work can be performed safely.

NOTES

NOTES

14

Safety Equipment and Rescue

United States Occupational Safety and Health Administration (OSHA) Safety Rules

- OSHA regulation (standards—29 CFR) 1910.269, *Electric Power Generation, Transmission, and Distribution, Subpart R—Special Industries,* covers the operation and maintenance of electric power generation, control, transformation, transmission, and distribution lines and equipment.

National Electrical Safety Code

Per the standards abstract: "This standard covers basic provisions for safeguarding of persons from hazards arising from the installation, operation, or maintenance of

- Conductors and equipment in electric supply stations, and
- Overhead and underground electric supply and communication lines.

It also includes work rules for the construction, maintenance, and operation of electric supply and communication lines and equipment."

National Electrical Code®

This standard's purpose is to safeguard "persons and property from hazards arising from the use of electricity." The NEC® is typically for electrical equipment and systems connected on the load side of an electric utility's electric meter. For electric utilities, applicability is very well defined in Article 90, Scope 90-2 in that electrical overhead and underground equipment controlled exclusively by the public utility for the generation, transmission, and distribution of electricity defers to the rules and regulations in the NESC. However, the utility's customer offices are expected to meet requirements contained in the NEC.

De-energized Work

The safest way of completing electric distribution work is to de-energize, isolate, test, and ground the facilities to be worked on. The lineman must correctly identify the circuit to be grounded and obtain from the proper authority assurances that the circuit has been de-energized and isolated.

Hold-off tags must be applied to the switches isolating the circuit, and these must be identified with the name of the person in charge of the work. A high-voltage tester mounted on hot-line tools can be used to verify that the circuit is de-energized. The grounding conductors are first connected, to a low-resistance ground, usually the neutral conductor, and then to the de-energized phase conductors by the lineman with a hot-line tool. Once the work is completed and all personnel are clear of the circuit, the grounds are removed in the reverse sequence.

Hot-Line Work

The public's dependence on continuous electric service to maintain their health, welfare, and safety often makes it impossible to de-energize circuits to complete modifications and perform maintenance. Proper application of the protective equipment is necessary for the lineman to safeguard the work area, that is, to prevent electric shock or flash burn injuries. Arc flash safety assessments and their need (arc exposures > 2 cal/cm^2) are defined and addressed within the 2007 NESC. While the elimination of arc flash hazards is desirable, some arc flash hazards may exist. The hazards are to be understood and the proper safety measures and programs in place for everyone in the vicinity. Hot-line work is performed while men are standing in or on an insulated platform or from aerial lift devices. Low primary voltages may be worked directly from the pole by linemen. The higher primary voltages must be worked from an insulated platform attached to the pole or an insulated aerial device. The work area is guarded by conductor covers, insulated covers, crossarm covers, rubber blankets, and dead-end covers. The energized conductors have been elevated and supported clear of the crossarm with hot-line tools.

Protective Equipment

Linemen and cablemen must use protective equipment to safely complete work on energized equipment. Correct application of the protective equipment described is essential for injury prevention and the maintenance of reliable electric service. Linemen wear: hard hats designed for electrical work, safety glasses, body belts with safety straps, rubber gloves with protectors, and rubber sleeves.

Rubber gloves

■ The most important article of protection for a lineman or a cableman is a good pair of rubber gloves (see Table 14.1) with the proper dielectric strength for the voltage of the circuit to be worked on.

■ Leather protector gloves must always be worn over the rubber gloves to prevent physical damage to the rubber while work is being performed.

■ When the rubber gloves are not in use, they should be stored in a canvas bag to protect them from mechanical damage or deterioration from ozone generated by the sun.

■ Rubber gloves should always be given an air test by the lineman or cableman each day before the work is started or if the workman encounters an object that may have damaged the rubber gloves.

■ The American National Standards Institute standard ANSI/ASTM D 120, "Rubber Insulating Gloves," defines lineman's rubber glove specifications.

■ The proof-test voltage of the rubber gloves should not be construed to mean the safe voltage on which the glove can be used.

■ The maximum voltage on which gloves can safely be used depends on many factors including the care exercised in their use; the care followed in handling, storing, and inspecting the gloves in the field; the routine established for periodic laboratory inspection and test; the quality and thickness of the rubber; the design of the gloves; and such other factors as age, usage, or weather conditions.

TABLE 14.1   Classes of Rubber Gloves Manufactured, Proof-Test Voltages, and Maximum-Use Voltages as Specified by ANSI/ASTM D 120-77

Class of glove	AC proof-test voltage, rms	DC proof-test voltage, average	Maximum-use voltage, AC rms
0	5,000	20,000	1,000
1	10,000	40,000	7,500
2	20,000	50,000	17,000
3	30,000	60,000	26,500
4	40,000	70,000	36,000

Reprinted by permission of the American Society for Testing and Materials, 1916 Race Street, Philadelphia, PA 19103. Copyright.

- Inasmuch as gloves are used for personal protection and a serious personal injury may result if they fail while in use, an adequate factor of safety should be provided between the maximum voltage on which they are permitted to be used and the voltage at which they are tested.

- It is common practice for the rubber protective equipment to prepare complete instructions and regulations to govern in detail the correct and safe use of such equipment. These should include provision for proper-fitting leather protector gloves to protect rubber gloves against mechanical injury and to reduce ozone cutting. The lineman wear rubber gloves with leather protectors and rubber insulating sleeves.

- Rubber insulating gloves should be thoroughly cleaned, inspected, and tested by competent personnel regularly.

A procedure for cleaning, inspecting, and testing rubber gloves is as follows:

1. Rubber labels showing the glove size, the glove number, and the test date are cemented to each rubber glove.

2. The rubber gloves are washed in a washing machine, using a detergent to remove all dirt, prior to testing.

3. The rubber gloves are inflated by utilizing glove inflator equipment, checked for air leaks, and inspected. The inspection includes checking the surface of the rubber gloves for defects such as cuts, scratches, blisters, or embedded foreign material. The cuff area of the inside of the gloves is visually inspected after the glove is removed from the inflator. All rubber gloves having surface defects are rejected or repaired.

4. After the inflated inspection, the rubber gloves are stored for 24 hours to permit the rubber to relax, thus minimizing the possibility of corona damage while the electrical tests are completed.

5. The rubber gloves are supported in a testing tank by means of plastic clothespins on the supporting rack. A ground electrode is placed inside each glove.

6. The rubber gloves to be tested and the testing tank are filled with water. The height of the water in the glove and in the testing tank depends upon the voltage test to be applied.
   Class 0 gloves are tested at 5000-VAC or 20,000-VDC. The flashover clearances at the top edge of the cuff of the gloves during the test must be $1\frac{1}{2}$ inches for both the AC and DC test.
   Class 1 gloves are tested at 10,000-VAC or 40,000-VDC. The flashover clearances at the top edge of the cuff of the gloves during the test must be $1\frac{1}{2}$ inches for the AC test and 2 inches for the DC test.

Class 2 gloves are tested at 20,000-VAC or 50,000-VDC. The flashover clearances at the top edge of the cuff of the gloves during the test must be 2¹/₂ inches for the AC test and 3 inches for the DC test. Class 3 gloves are tested at 30,000-VAC or 60,000-VDC. The flashover clearances at the top edge of the cuff of the gloves during the test must be 3¹/₂ inches for the AC test and 4 inches for the DC test. Class 4 gloves are tested at 40,000-VAC or 70,000-VDC. The flashover clearances at the top edge of the cuff of the gloves during the test must be 5 inches for the AC test and 6 inches for the DC test.

7. Sphere-gap setting for the test voltage to be applied (Table 14.2). Sphere-gap settings will change with variations of temperature and humidity.

8. Connect the high-voltage lead of the testing transformer to the metal tank portion of the rubber-glove-testing equipment.

9. Connect the ground electrode in each rubber glove to its associated ground lead.

10. Evacuate all personnel from the enclosed testing area.

11. Close the power-supply-isolating switch to the testing equipment.

12. By means of the remote controls, gradually raise the voltage output of the testing transformer from zero to the proper level at a rate of approximately 1000 V/sec. This voltage should be applied to the rubber gloves for a period of 3 minutes.

13. After the high voltage has been applied to the rubber gloves for 3 min, switch the milliammeter to read the leakage current in each ground electrode utilized for each rubber glove.

14. The output voltage of the test transformer is gradually reduced to zero at a rate of approximately 1000 V/second.

15. All rubber gloves that failed during the test or had a current leakage greater than the maximum permissible value are rejected (Table 14.3). In some instances the gloves can be repaired and used if they successfully pass a retest.

TABLE 14.2   Sphere-Gap Settings

Test voltage AC, volts	Sphere-gap setting, mm
5,000	2.5
10,000	5.0
20,000	10.0
30,000	15.0
40,000	22.0

TABLE 14.3   Maximum Leakage Current for Five Classes of Rubber Gloves

Class of glove	Test voltage, AC	Maximum leakage current, milliamps			
		10½ in*	14 in*	16 in*	18 in*
Class 0	5,000	8	12	14	16
Class 1	10,000	–	14	16	18
Class 2	20,000	–	16	18	20
Class 3	30,000	–	18	20	22
Class 4	40,000	–	–	22	24

Reprinted by permission of the American Society for Testing and Materials, 1916 Race Street, Philadelphia, PA 19103. Copyright.
*Glove cuff length

16. All rubber gloves which pass the high-voltage test are dried on a drying rack and powdered, and a tag indicating the leakage current recorded for each glove is placed in each glove. The gloves are then returned to the linemen or cablemen for use. In-service care of insulating gloves is specified in ANSI/ASTM F 496-77.

Rubber sleeves

■ Rubber sleeves should be worn with rubber gloves to protect the arms and shoulders of the lineman, while he is working on high-voltage distribution circuits, from electrical contacts on the arms or shoulders. Rubber insulating sleeves must be treated with care and inspected regularly by the linemen in a manner similar to that described for rubber insulating gloves.

■ The rubber insulating sleeves should be thoroughly cleaned, inspected, and tested by competent personnel regularly. The inspecting and testing procedures are similar to those described for rubber insulating gloves.

■ In-service care of insulating sleeves is specified in ANSI/ASTM F 496.

Rubber insulating line hose

■ Primary distribution conductors can be covered with rubber insulating line hose to protect the lineman from an accidental electrical contact.

■ The line hoses are manufactured in various lengths, with inside-diameter measurements that vary from 1 to 1½ inches and with different test voltage withstand values.

- The lineman should be sure that the voltage rating of the line hose provides an ample safety factor for the voltage applied to the conductors to be covered.

- All line hoses should be cleaned and inspected regularly. A hand-crank wringer can be used to spread the line hose to clean and inspect it for cuts or corona damage.

- The rubber insulating line hose should be tested for electric withstand in accordance with the specifications at scheduled intervals.

- In-service care of insulating line hose and covers is specified in ANSI/ASTM F 478.

Rubber insulating insulator hoods

- Pin-type or post-type distribution primary insulators can be covered by hoods.

- The insulator hood, properly installed, will overlap the line hose, providing the lineman with complete shielding from the energized conductors.

- Insulator hoods, like all other rubber insulating protective equipment, must be treated with care, kept clean, and inspected at regular intervals.

- Canvas bags of the proper size attached to a hand line should be used to raise and lower the protective equipment when it is to be installed and removed.

Conductor covers

- A conductor cover, fabricated from high-dielectric polyethylene, clips on and covers conductors up to 2 inches in diameter.

- A positive air gap is maintained by a swinging latch that can be loosened only by a one-quarter turn with a clamp stick.

Insulator covers

- Insulator covers are fabricated from high-dielectric polyethylene and are designed to be used in conjunction with two conductor covers.

- The insulator cover fits over the insulator and locks with a conductor cover on each end.

- A polypropylene rope swings under the crossarm and hooks with a clamp stick, thus preventing the insulator cover from being removed upward by bumping or wind gusts.

Crossarm covers

- High-dielectric-strength polyethylene crossarm covers are used to prevent tie wires from contacting the crossarm when conductors adjacent to insulators are being tied or untied.

- It is designed for single- or double-arm construction, with slots provided for the double-arm bolts.

- Flanges above the slots shield the ends of the double-arm bolts.

Pole covers

- Polyethylene-constructed pole covers are designed to insulate the pole in the area adjacent to high-voltage conductors.

- The pole covers are available in various lengths.

- Positive-hold polypropylene rope handles are knotted through holes in the overlap area of the cover.

Rubber insulating blankets

- Odd-shaped conductors and equipment can usually be covered best with rubber insulating blankets.

- The rubber insulating blankets are stored in canvas rolls or metal canisters to protect them when they are not in use. The blankets can be held in place by ropes or large wooden clamps.

- In-service care of insulating blankets is specified in ANSI/ASTM F 479.

Safety hat

- Hard hats, or safety hats, are worn by linemen, cablemen, and groundmen to protect the worker against the impact from falling or moving objects and against accidental electrical contact of the head and energized equipment, as well as to protect the worker from the sun, cold, rain, sleet, and snow.

- The first combined impact-resisting and electrical insulating hat was introduced in 1952.

- The hat was designed "to roll with the punch" by distributing the force of a blow over the entire head.

- This is accomplished by a suspension band holding the hat about an inch away from the head and letting the hat work as a shock absorber.

- The hat is made of fiberglass, or plastic material, and has an insulating value of approximately 20,000 V.

- New helmets are manufactured to withstand a test of 30,000 volts without failure.

- The actual voltage that the hat will sustain while being worn depends upon the cleanliness of the hat, weather conditions, the type of electrode contacted, and other variables.

- The wearing of safety hats by linemen and cablemen has greatly reduced electrical contacts.

- Physical injuries to the head have been practically eliminated as a result of workers on the ground wearing protective helmets.

- The *Occupational Safety and Health Act of 1970* and most companies' safety rules require linemen, cablemen, and groundmen to wear safety hats while performing physical work.

- Specifications for safety hats are found in ANSI Standard Z89.2, *Safety Requirements for Industrial Protective Helmets for Electrical Workers, Class B.*

Resuscitation

This section will present the steps you should take in an emergency by offering a review and explanation of what you have learned in professionally given first aid, CPR and pole-top rescue classes. It is not intended as a replacement for taking the *OSHA*-required American Red Cross, the American Heart Association, the U.S. Bureau of Mines, or an equivalent training that can be verified by documentary evidence.

The procedures in this section are based on the most current recommendations available at the time of printing. We make no guarantees as to and assume no responsibility for the correctness, or completeness of the presented information or recommendations.

To be prepared for an emergency, you should take a review course yearly and practice the skills on a mannequin in the presence of a trained instructor. Techniques are being updated, and you need to learn the current skills. The American Red Cross, American Heart Association, the U.S. Bureau of Mines, and others have courses which you may take.

Be prepared for an emergency. The best way to care for an emergency is to PREVENT IT. THINK SAFETY.

- Have and use protective equipment.

- Inspect rubber gloves daily and have them electrically tested, as per industry recommendations,

- Be aware of what is going on around you.

- Keep your mind on what you are doing.

- Do not use any substance that will impair your ability to act and to react.

- Avoid wearing the orange traffic vest when not directing traffic.

- Wear clothing that is nonconductive and flame resistant.

First-aid equipment

OSHA 1926.50 defines the medical services and first aid which should be available to the employee.

Prompt medical attention should be available within a reasonable access time (4 minutes). If not available, a person who has a valid first aid, CPR certificate must be at the work site to give first aid.

First-aid kits and supplies should be in a weatherproof container with individual sealed packages for each type of item. The contents shall be checked by the employer before being sent out on each job and at least weekly thereafter to see that expended items are replaced.

First-aid kits are for your use and protection on the job. When they are taken off of the truck for unauthorized use and an emergency occurs, the missing supplies could mean the difference between an incident of small proportions and a life-threatening incident.

The employer is responsible for the first-aid kits and so are you. You need to make sure they are present, the supplies are not out of date or soiled. Make sure the appropriate company personnel knows that supplies need to be replaced.

This is your responsibility. Your life could depend upon it.

What to do in an emergency

- Be prepared.

- Remain calm.

- If the victim is on a pole or in a bucket, shout at your victim to see if they are conscious and can respond.

- If they cannot respond, immediately send someone to call 911 or your local emergency system.

- Use common sense.

- We do not want two victims.

- Do not approach the victim until it is safe for you to do so.

- Assess your victim.

- Use protective barriers.

- Give rescue breathing, CPR or Heimlich maneuver as needed (these skills will be reviewed for you in the next pages).

Effects of electric shock on the human body

The effect of electric shock on a human being is rather unpredictable and may manifest itself in a number of ways.

- *Asphyxia.* Electric shock may cause a cessation of respiration (asphyxia). Current passing through the body may temporarily paralyze (or destroy) either the nerves or the area of the brain which controls respiration.

- *Burns (contact and flash).* Contact burns are a common result of electric current passing through the body. The burns are generally found at the points where the current entered and left the body and vary in severity, the same as thermal burns. The seriousness of these burns may not be immediately evident because their appearance may not indicate the depth to which they have penetrated. In some accidents there is a flash or electric arc, the rays and heat from which may damage the eyes or result in thermal burns to exposed parts of the body.

- *Fibrillation.* Electric shock may disturb the natural rhythm of the heartbeat. When this happens, the muscles of the heart are thrown into a twitching or trembling state, and the actions of the individual muscle fibers are no longer coordinated. The pulse disappears, and circulation ceases. This condition is known as *ventricular fibrillation* and is serious.

- *Muscle spasm.* A series of erratic movements of a limb or limbs may occur owing to alternating contractions and relaxations of the muscles. This muscle spasm action on the muscles of respiration may be a factor in the stoppage of breathing.

Rescue

Because a person may receive electric shock in many different locations (on the ground, in buildings, on poles, or on steel structures), it is neither possible nor desirable to lay down definite methods of rescue. However, certain facts should be remembered.

Freeing the victim

Because of the muscle spasm at the time of shock, most victims are thrown clear of contact. However, in some instances (usually low voltage) the victim is still touching live equipment. In either situation, the rescuer must be extremely careful not to get in contact with the live equipment or to touch the victim while he or she is still in contact. The rescuer should "free" the victim as soon as possible so that artificial respiration can be applied without hazard. This may involve opening switches or cutting wires so that equipment within reach is de-energized or using rubber gloves or other approved insulation to move the victim out of danger.

If the victim is to be lowered from a pole, tower, or other structure, a hand line of adequate strength, tied or looped around him and placed over a crossarm or tower member, is a simple and satisfactory method.

Artificial respiration methods

Early methods of artificial respiration involved applying heat by building a fire on the victim's stomach, beating the victim with a whip, inverting the victim by suspending him by his feet, or rolling the victim over a barrel. Modern methods of artificial respiration include the Schafer prone method, the chest-pressure-arm-lift method, the back-pressure-arm-lift method, and the mouth-to-mouth method. The mouth-to-mouth method is currently advocated by the Edison Electric Institute and the American Red Cross. The mouth-to-mouth method accepted by most utility companies is depicted in Fig. 14.1. Artificial respiration should be used when breathing stops as a result of electric shock, exposure to gases, drowning, or physical injury. If the heart has stopped functioning as a result of cardiac arrest or ventricular fibrillation, heart-lung resuscitation should be used with mouth-to-mouth resuscitation.

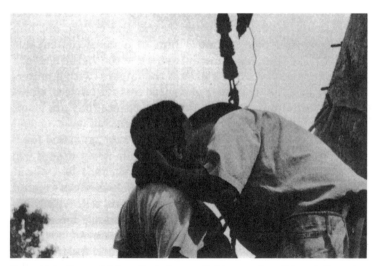

Figure 14.1 Two quick breaths being applied to a nonbreathing victim prior to lowering to the ground.

Applying artificial respiration

To be successful, artificial respiration must be applied within the shortest time possible to a victim who is not breathing. Without oxygen, the brain begins to die. The first 4 minutes are critical. Irreversible brain damage begins. The best chance of survival is for the victim to receive rescue breathing, CPR immediately and advanced medical care from an Emergency Medical Technician (EMT).

Normally, the type of resuscitation in which the victim is placed on the ground or floor is the best because the victim can be given additional treatment (stoppage of bleeding, wrapping in blanket to keep warm, etc.) and the rescuer can be easily relieved without an interruption of the breathing rhythm or cycle. However, pole-top resuscitation (or any adaption of it) should be started if there will be any appreciable delay in getting the victim into the prone position.

In choosing the type of resuscitation to be used, the rescuer must consider the obvious injuries suffered by the victim. Broken ribs, for example, might make inadvisable the use of certain types of resuscitation; burns on the arms or on the face might exclude other types. However, no time should be lost in searching for injuries; artificial respiration should be started at once.

The need for speed

Artificial respiration must be started at once. Time is the most important single factor in successful artificial respiration. The human brain can exist for only a few minutes, probably not more than four or five, without oxygenated blood; therefore, every second counts. The greater the delay, the smaller the chance of successful recovery. There should be no delay to loosen clothing, warm the victim, get him down from the pole, or move him to a more comfortable position. However, an immediate check of the victims mouth should be made by a quick pass of the fingers (with a protective barrier such as rubber gloves) to remove tobacco, chewing gum, etc. Do not remove dentures unless they are loose. They help to support the facial structure.

Use protective barriers

In today's society, we run the risk of ourselves being exposed and exposing the victim to diseases such as HIV/AIDS, TB, hepatitis, viral infections, and some forms of pneumonia. Use protective barriers when possible.

Resuscitation masks vary in size, shape, and features. The most effective are ones made of transparent, pliable material that allows you to make a tight seal on the person's face and have a one-way valve for releasing a person's exhaled air. Use rubber gloves when available.

Assess your victim

You must act immediately. Send someone to call 911 or your local emergency system immediately. Determine and use the best method of rescuing the victim. If the injured person is on a pole or in a bucket, make sure the area is safe for you to approach and then begin pole-top-or bucket-rescue procedures.

Shout at the victim on the pole to see if he can respond and is conscious. If he cannot respond, and the area is safe to reach the victim, determine if your victim is breathing. If not breathing, give two rescue breaths before you lower him to the ground. Use protective barriers if possible.

Lower victim to ground, open the airway, and determine if victim is breathing. To open the airway, you should tilt the head back by placing one hand on the victim's forehead. Using your other hand, place two fingers under the bony part of the lower jaw near the chin. Tilt the head back and lift the jaw. Avoid closing the victim's mouth. Avoid pushing on soft parts under the chin.

The head-tilt/chin-lift will remove the tongue away from the back of the throat. Many times a victim will begin breathing on his or her own once you have done this.

If not breathing, give two rescue breaths.   A rescue breath is one breath given slowly for approximately 1 1/2 to 2 seconds. If you find that air will not enter or that the chest is not rising and falling, you should reposition the head and attempt giving a breath. Improper chin and head positioning is the most common cause of difficulty with ventilations. If after you have repositioned the airway and air will not go in, you may have a foreign object trapped in the airway. You should begin airway obstruction maneuver (Heimlich maneuver).

While maintaining the head-tilt/chin-lift, the rescuer should take a deep breath and seal their lips around the victim's mouth, creating an airtight seal. The rescuer gently pinches the nose closed with the thumb and index finger of the hand on the forehead, thereby preventing air from escaping through the victim's nose.

Gently blow into the victim's mouth for 1 1/2 to 2 seconds. Look for the chest to rise. Release the nose so that the air can escape. Pause only long enough for you to take a breath, then repeat the procedure for a second rescue breath.

Check for pulse

After you have given two rescue breaths, check for pulse at the carotid artery. Feel the pulse for approximately 5 to 10 seconds. If pulse is present, but the victim is not breathing, begin rescue breathing. If pulse is absent, begin CPR.

Mouth-to-nose rescue breathing

The *Journal of the American Medical Association* recommends that mouth-to-nose breathing be used when it is impossible to ventilate through the victim's mouth, the mouth cannot be opened, the mouth is seriously injured, or a tight mouth-to-mouth seal is difficult to achieve.

Keep the victim's head tilted back with one hand on the forehead and use the other hand to lift the victim's lower jaw (as in the head-tilt/chin-lift) and close the mouth. The rescuer takes a deep breath and then breathes into the victim's nose. The rescuer removes his mouth and allows the air to escape.

If a pulse is present, but the person is not breathing, begin rescue breathing

1. Give the victim one rescue breath about every 5 seconds.

2. Do this for about 1 minute (12 breaths).

3. Recheck the pulse and breathing about every minute.

4. You should continue rescue breathing as long as a pulse is present and the victim is not breathing, or until one of the following things happen:
   - The victim begins to breathe without your help.
   - The victim has no pulse (begin CPR).
   - Another trained rescuer takes over for you.
   - You are too tired to continue.

Review

1. Recognize that you have an emergency.

2. Call for help immediately.

3. Assess your victim for unresponsiveness.

4. Be sure that the scene is safe.

5. Begin pole-top or bucket rescue.

6. Check for breathing, if not give two breaths (remember use protective barrier, if available)

7. Check for pulse.

8. If pulse is present and breathing is absent, begin rescue breathing (one breath every 5 seconds).

9. If there is no pulse, start CPR.

Airway obstruction: conscious adult

Ask the victim if he or she is choking and if you can help him or her. If the victim is coughing forcefully, the rescuer should not interfere with the victim's own attempt to expel the foreign object, but stay with the person and encourage him or her to continue coughing (Fig. 14.2).

If the victim has a weak, ineffective cough or is making high-pitched noises while inhaling or is unable to speak, breathe, or cough, this

Figure 14.2   Universal sign for choking.

should be treated as a complete airway obstruction. Stand behind the victim. Place your leg between the victim's legs (this will help you control the victim if she or he passes out).

Wrap your arms around the victim's waist, make a fist with one hand, the thumb side of the fist is placed against the victim's abdomen just slightly above the navel and well below the tip of the breastbone (Fig. 14.3).

The fist is grasped with the other hand and pressed into the victim's abdomen with a quick upward thrust. Each new thrust should be a separate and distinct movement administered with the intent of relieving the obstruction. The thrusts should be repeated and continued until the object is expelled from the airway or the patient becomes unconscious.

Figure 14.3   A quick upward thrust being applied to the choking victim.

Conscious adult choking victim
who becomes unconscious

If the victim becomes unconscious, lower to the floor or ground. The rescuer straddles the victim and places the heel of one hand slightly above
the navel and well below the tip of the breastbone. The second hand is
placed directly on top of the fist. The rescuer presses into the abdomen
with quick upward motion.

Give up to five abdominal thrusts. The victim's mouth should be
opened and a finger sweep done after you have given five abdominal
thrusts (Heimlich maneuver). Give the victim a rescue breath; if air does
not go in, repeat the sequence of five abdominal thrusts, finger sweep,
and attempt to ventilate. Repeat as long as necessary.

How to do a finger sweep

A finger sweep is done by opening the mouth; grasp both the tongue and
the lower jaw between the thumb and fingers of one hand and lift jaw.
Insert index finger of other hand into the mouth, running it along the
inside of the cheek and deep into throat to the base of the tongue. Use a
"hooking" action to dislodge any object that might be there. Remove the
object. Be careful not to force the object deeper into the airway.

What to do when the object is cleared

If the object is cleared and the victim is conscious, the victim may want
to leave the area and go to the restroom, etc.

Do not let them leave by themselves. The throat has been irritated and
may swell and the airway may close. Stay with the victim until EMS personnel arrive.

If you are alone and choking

If you are alone and choking, you can perform the Heimlich maneuver
on yourself. If this is unsuccessful, the upper abdomen should be pressed
quickly over any firm surface such as the back of a chair, side of a table,
or porch railing. Several thrusts may be needed to clear the airway.

What to do if your victim is pregnant or large

If you are not able to get your arms around the waist of a choking victim
who is conscious, you must give chest thrusts.

Chest thrusts are given by getting behind the victim, (the victim can
be sitting or standing). Place your arms under the victim's armpits and
around the chest. Place the thumb side of your fist on the middle of the
breastbone. Be sure that your fist is centered right on the breastbone

and not on the ribs. Make sure that your fist is not near the lower tip of the breastbone.

Grasp your fist with your other hand and give backward thrusts. Give thrusts until the obstruction is cleared or until the person loses consciousness. You should think of each thrust as a separate attempt to dislodge the object.

Review

1. Ask if victim is choking.

2. Ask for permission to help.

3. If victim is coughing forcefully, encourage her or him to continue coughing.

4. If victim has obstructed airway and is conscious, do abdominal thrusts until object is expelled or victim becomes unconscious.

5. If victim becomes unconscious, lower her or him to the floor.

6. Give five abdominal thrusts.

7. Do a finger sweep.

8. Give a rescue breath; if air still does not go in, repeat sequence of five thrusts, finger sweep, and breath.
   - If alone, give Heimlich maneuver (abdominal thrust) to yourself, or lean over firm surface.

Heart-lung resuscitation: cardiopulmonary resuscitation (CPR)

1. Kneel facing the victim's chest. Find the correct hand position by sliding your fingers up the rib cage to the breastbone.

2. Place your middle finger in the notch and the index finger next to it on the lower end of the breastbone.

3. Place your other hand beside the two fingers.

4. Place hand used to locate notch on top of the other hand. Lace the fingers together. Keep fingers off of the chest. Position shoulders over hands, with elbows locked and arms straight.

Hands vary in size and shape. Arthritic or injured hands and wrists will not allow some rescuers to lace their fingers together. An acceptable alternative hand position is to grasp the wrist of the hand on the chest with the hand that has been locating the lower end of their sternum. Another acceptable alternative is to hold your fingers up off the chest. Use the method that works best for you. Just remember to keep your fingers off the chest. Use the heel of your hand to do the compressions.

1. Compress the breastbone 1^{1}/$_{2}$ to 2 inches.

2. Give 30 compressions. Count 1 and 2 and 3 and 4 .... Give two rescue breaths after the 30th compression. The cycle of 30 compressions and two breaths is consistent with ensuring the blood continues circulation within the body.

3. Do four cycles of 30 compressions and two breaths. Check the pulse at the end of the fourth cycle.

When you press down, the weight of your upper body creates the force you need to compress the chest. It is important to keep your shoulders directly over your hands and your elbows locked so that you use the weight of your upper body to compress the chest. Push with the weight of your upper body to compress the chest. Push with the weight of your upper body, not with your arm muscles. Push straight down; don't rock.

If the thrust is not in a straight downward direction, the torso has a tendency to roll; part of the force is lost, and the chest compression may be less effective.

Once you have started CPR, you must continue until one of the following happen:

- Another trained person takes over CPR for you.

- EMS personnel arrive and take over the care of the victim.

- You are too exhausted and unable to continue.

- The scene becomes unsafe.

Review

1. Call for advance help (911).

2. Determine that victim has no pulse.

3. Find correct hand compression.

4. Do four cycles of 30 compressions and two breaths.

5. Recheck pulse for 5 seconds.

6. If pulse is not present, continue with CPR.

Why do I do that and why did they change?

1. *We used to give CPR for a minute and then go call for help. Why has it been changed to "Call for help immediately and then begin CPR?"*

It was found that trained single rescuers often performed CPR much longer than one minute. Thereby, the call to an EMS system or an ACLS care was seriously delayed. It is important for the victim to receive the advanced life-saving medical support as soon as possible.

2. *Why do we use the head-tilt/chin-lift now instead of the old way of pushing up on the neck?*

Using the head-tilt/chin-lift method helps to minimize further damage to a victim. It is more efficient in opening and sealing the airway.

3. *Why do we ask if a victim is choking when it appears obvious that they are?*

An allergic reaction can also cause the airway to close. In such a case, the victim needs advanced medical support immediately and all of the abdominal thrusts in the world is not going to open the airway.

4. *What if the jaw or mouth is injured, the mouth is shut too tight, or your mouth is too small?*

You should do mouth-to-nose breathing.

5. *What if a victim is breathing on his own but is unconscious?*

It is vital that professional emergency personnel be called immediately. Victims who may be breathing on their own may still go into shock if counteractive measures are not begun immediately. There is a point when shock becomes irreversible and the victim will die.

6. *Why has the time taken during ventilation for filling the lungs of an adult increased to $1^{1}/_{2}$ to 2 seconds per breath?*

The increased time further decreases the likelihood of the stomach filling with the air and the victim then vomiting.

7. *What's the big deal if a victim vomits?*

When an unconscious victim vomits, stomach contents may get into the lungs, obstructing breathing, which may hamper rescue breathing attempts, thus being fatal. Residue in the lungs can also cause complications after breathing has been restored.

8. *Why do we no longer give hard breaths, but slow breaths?*

By giving slow breaths, you can feel when the lungs are full and you meet resistance. Over inflating the lungs can cause air to go into the stomach.

9. *Do we remove dentures?*

No. The dentures should be removed only if they cannot be kept in place.

10. *What if the victim has something in his or her mouth like chewing gum, tobacco, or vomit?*

If foreign material or vomit is visible in the mouth, it should be removed quickly. Liquids or semiliquids should be wiped out with the index and middle fingers. The fingers should be covered by a protective barrier such as rubber gloves or by a piece of cloth. Solid material should be extracted with a hooked index finger. Remember to use *protective barriers whenever possible.*

11. *What causes respiratory arrest?*

Many things can cause respiratory arrest. Shock, allergic reactions, drowning, and electrocution.

12. *Why is it imperative that two rescue breaths be given before you bring the victim down from the pole?*

Extensive soft-tissue swelling may develop rapidly and close the airway. The heart may stop beating soon after breathing has stopped. The heart will continue to beat in some instances after breathing has stopped. Giving rescue breathing before bringing the victim down helps delay the heart stopping and brain damage from lack of oxygen (Fig. 14.4).

13. *Why do you stop breathing after an electrocution?*

In electrocution, depending upon the path of the current, several things may occur. Your victim may continue to have a heart rhythm, but not be breathing because of

- The electrical current having passed through the brain and disrupting the area of the brain that controls breathing.

Figure 14.4   A victim being lowered from the pole.

- The diaphragm and chest muscles which control breathing may have been damaged.
- The respiratory muscles may have been paralyzed.

If artificial respiration is not started immediately, the heart will stop beating.

14. *I know CPR and I am good at my skills, why was I not able to save my victim? Did I do something wrong?*

There are some instances when all the skills you possess will not be able to save the victim. With your skills, they had a chance, whereas if you did nothing, the victim had no chance. A hand-to-hand pathway is more likely to be fatal than a hand-to-foot or a foot-to-foot.

Electricity in the body causes organs (the heart, etc.) and bones to explode, rupture, and break as well as tissue destruction. When the organ has ruptured, there is very little if anything you would have been able to do. This is why it is so important that you use safe work procedures and protective equipment to prevent accidents.

15. *How do I know that I am giving adequate ventilations?*
   a. You should be able to observe the rise and fall of the chest with each breath you give.
   b. You should be able to hear and feel the escape of air during the passive exhalation of your victim.

16. *What causes a victim to vomit when they are not even breathing?*
   a. During rescue breathing, you may not have the airway opened properly. You are forcing air into the stomach and not into the lungs.
   b. The breath you give is given too long, at too fast or at too much pressure. The air is forced into the stomach.

17. *Is it a problem if the stomach is full of air?*
   a. A stomach filled with air reduces lung volume by elevating the diaphragm.
   b. The victim may also have the stomach contents forced up by the air in the stomach.

18. *What do I do if the victim starts to vomit?*

Turn the victim onto one side, wipe out his or her mouth with a protective barrier. Reassess your victim to see if they are breathing on their own. If not breathing, continue with the rescue procedure you are using (i.e., rescue breathing, CPR). If you are alone and must leave an unconscious victim, position the victim on his or her side in case the victim vomits while you are gone.

19. *How can someone get enough oxygen out of the air we exhale? Didn't I use all of the oxygen?*

The air we breathe has approximately 21% oxygen. We use about 5% of the oxygen. That means that we exhale air containing approximately 16% of oxygen. This shows you that there is enough oxygen in the air you breath out to support life until advanced medical care arrives.

20. *Do chest compressions circulate the same amount of blood as the heart does normally?*

Under the best of conditions, chest compressions only circulate approximately 1/3 of the normal blood flow. Body tissues are receiving only the barest minimum of oxygen required for short term survival. That is why it is so important to get professional help immediately.

21. *What does CPR stand for and what does it do?*

CPR stands for *cardiopulmonary resuscitation. Cardio* refers to the heart and *pulmonary* refers to the lungs. CPR is a combination of chest compressions and rescue breathing. The purpose of CPR is to keep the lungs supplied with oxygen when breathing has stopped, and to keep blood circulating with oxygen to the brain, heart, and other parts of the body.

22. *Do we ever slap an adult choking victim on the back?*

No. You could lodge the item more securely.

23. *Why does a victim who is going to be given CPR need to be on a level surface?*

When the head is elevated above the heart, blood flow to the brain is further reduced or eliminated.

24. *What do I do if my hands come off of the victim's chest or move from the correct compression point when I am giving CPR?*

You must go up the rib cage and find the correct hand position again.

25. *Does the victim's chest need to be bared to perform compressions?*

No. Clothing should remain on unless it interferes with finding the proper location for chest compressions. Do not waste time or delay the starting of the compressions by removing clothing.

26. *Can or should you do CPR on someone who has a pacemaker?*

Yes. The pacemaker is placed to the side of the heart and directly below the breastbone, it will not get in the way of chest compressions.

27. *Do I give CPR to a victim who has a very-slow or weak pulse?*

No. Never perform CPR on a victim who has a pulse, it can cause serious medical complications.

28. *What if a victim regains a pulse and resumes breathing during or following resuscitation?*

The rescuer should continue to help maintain an open airway. The rescuer should then place the patient in the recovery position. To place the victim in the recovery position, the victim is rolled onto his or her side so that the head, shoulders, and torso move simultaneously without twisting. If the victim has sustained trauma or trauma is suspected, the victim should not be moved.

29. *How hard do I press when doing the head-tilt/chin-lift maneuver?*

The fingers must not press deeply into the soft tissue under the chin, put them on the bone. Pressing into the soft tissue might obstruct the airway. The thumb should not be used for lifting the chin.

30. *If there are several people who can help the victim, should one of them take off the tool belt and hooks?*

Yes. Smoldering shoes, belts, and hooks should be removed to prevent further thermal damage and possible injury to the rescuer. *Never* remove adhered particles of charred clothing from the body. It will pull the skin off with it. If you are the only rescuer, do not take time to remove items unless it interferes with the ability to administer effective compressions. It is critical to begin CPR immediately.

Pole-Top and Bucket-Truck Rescue

Time is critical. You may have to help a worker on a pole reach the ground safely when he or she

- Becomes ill
- Is injured
- Loses consciousness

You must know

- When he needs help
- When and why time is critical
- The approved method of lowering

Basic steps in pole-top rescue

- Evaluate the situation.
- Provide for your protection.
- Climb to rescue position.
- Determine the injured's condition.

Then, if necessary

- Give first aid.
- Lower injured.
- Give follow-up care.
- Call for help, etc.
- Evaluate the situation.
- Call to worker on pole.
- If the worker does not answer or appears stunned or dazed.
- Prepare to go to the worker's aid.
- Time is extremely important.
- Provide for your protection.
- Your safety is vital to the rescue.

**Personnel tools and rubber gloves
(rubber sleeves, if required)**

Check

- Extra rubber goods?
- Live line tools?
- Physical condition of the pole?
- Damaged conductors, equipment?
- Fire on the pole?
- Broken pole?
- Hand line on pole and in good condition?

Climb carefully to the rescue position and position yourself

- To ensure your safety
- To clear the injured from hazard
- To determine the injured's condition
- To render aid as required
- To start mouth-to-mouth, if required
- To lower injured, if necessary

The best position will usually be slightly above the injured person.

Determine the injured's condition

He or she might be

- Conscious
- Unconscious but breathing
- Unconscious and not breathing
- Unconscious, not breathing, and heart stopped

If the injured is conscious

- Time may no longer be critical.
- Give necessary first aid on pole.
- Reassure the injured.
- Help him or her descend pole.
- Give first aid on ground.
- Call for help, if necessary.

If the injured is unconscious but breathing

- Watch him or her closely in case the breathing stops.
- Lower the injured to the ground.
- Give first aid on the ground.
- Call for help.

If the injured is unconscious and not breathing

- Provide an open airway.
- Give him two rescue breaths.
- If he responds, continue mouth-to-mouth until he is breathing without help.
- If he does not respond to the first two rescue breaths, check skin color, check for pupil dilation.
- If pupil contracts and color is good, continue mouth-to-mouth until he is breathing without help.
- Then, help victim descend pole.
- Watch him closely, his breathing may stop again.
- Give any additional first aid needed.
- Call for help.

- If pupil does not contract and skin color is bad, the heart has stopped.
- Prepare to lower him immediately.
- Give two rescue breaths just before lowering.
- Lower him to the ground.
- Start heart-lung resuscitation.
- Call for help.

The method of lowering an injured worker is

- Safe
- Simple
- Available

Equipment needs

- 1/2 inch hand line

Procedure

- Position hand line
- Tie the injured
- Remove the slack in hand line
- Take firm grip on fall line
- Cut injured's safety strap
- Lower the injured lineman

Rescuer

- Position hand line over the crossarm or other part of the structure.
- Position line for clear path to ground (usually the best position is 2 or 3 ft from pole).
- The short end of the line is wrapped around fall line twice. (Two wraps around fall line.)
- The rescuer ties hand line around victim's chest using three half hitches.

Tying the injured

- Pass line around the injured's chest.
- Tie three half-hitches.

- Place the knot in front, near one arm pit, and high on chest.
- Snug the knot.
- Remove the slack in the hand line.
- With one rescuer, remove the slack on the pole.
- With two rescuers, the lineman on the ground removes slack.

Important

- Give two quick breaths, if necessary.
- Grip the fall line firmly.
- With one rescuer, hold the fall line with one hand.
- With two rescuers, the lineman on the ground holds fall line.
- Cut injured's safety strap.
- Cut strap on side opposite desired swing.
- Caution do not cut your own safety strap on the hand line.

Lowering the injured

- With one rescuer, guide the load line with one hand
- Control the rate of descent with your other hand.
- With two rescuers, the lineman on the pole guides the load line.
- The lineman on the ground controls the rate of descent.

(If a conscious lineman is being assisted in climbing down, the only difference is that enough slack is fed into the line to permit climbing freedom.)

One-man rescue versus two-men rescue

The rescues differ only in control of the fall line.
    Remember, the approved method of lowering an injured lineman is

- Position the hand line.
- Tie the injured.
- Remove slack in the hand line.
- Grip the fall line firmly.
- Cut the injured's safety strap.
- Lower the injured lineman.

If pole does not have a crossarm, rescuer places the hand line over the fiberglass bracket insulator support, or other substantial piece of equipment, (such as a secondary rack, neutral bracket, or guy-wire attachment), strong enough to support the weight of the injured worker. The short end of line is wrapped around the fall line twice, tied around fall line twice, and tied around the victim's chest using three half- hitches. The injured's safety strap is cut and the rescuer lowers injured lineman to the ground. The line must be removed and the victim eased on to ground.

Lay the victim on his or her back and observe if the victim is conscious. If the victim is conscious, time might no longer be critical. Give necessary first aid. Call for help.

If the injured is unconscious and not breathing, provide an open airway. Give two rescue breaths. If he or she responds, continue mouth-to-mouth resuscitation until the injured lineman is breathing without help. If no pulse is present, pupils do not contract, and the skin color is not normal, the heart has stopped. Restore circulation. Use heart-lung resuscitation.

Bucket-truck rescue

- Time is critical.

- Equip a portion of the insulated boom of the truck with rope blocks designed for hot-line work.

- A strap is placed around the insulated boom, approximately 10 ft from bucket to support the rope blocks.

- Blocks are held taut on the boom from the strap to the top of the boom.

- Rescuer on ground evaluates the conditions when an emergency arises.

- The bucket is lowered using the lower controls.

- Obstacles in the path of the bucket must be avoided.

- The hook on the rope blocks is engaged in a ring on the lineman's safety strap.

- The safety strap is released from the boom of the truck.

- Rope blocks are drawn taut by the rescuer on the ground.

- The unconscious victim is raised out of the bucket with rope blocks.

- The rescuer eases victim on to the ground.

- Care should be taken to protect the injured victim from further injury.

- Release the rope blocks from the victim.

- The basket or bucket on aerial device may be constructed to tilt after being released.

- The tilt release eliminates the need for special rigging to remove an injured lineman from the basket or bucket.

- Lay the victim on their back and observe if the victim is conscious.

- If the victim is conscious, time might no longer be crucial.

- Give necessary first aid.

- Call for help.

- If injured is unconscious and not breathing, provide an open airway.

- Give the victim two rescue breaths

- If he responds, continue mouth-to-mouth resuscitation until the victim is breathing without help.

- If no pulse is present, pupils do not contract, and skin color is not normal, the victim's heart has stopped.

- Restore circulation.

- Use heart-lung resuscitation
  - Mouth-to-mouth resuscitation = external cardiac compression = heart-lung resuscitation
  - After every 30 compressions give 2 breaths, ratio 30 to 2

- Continue heart-lung resuscitation until the victim recovers, reaches a hospital, or a doctor takes over.

Important

Continue heart-lung resuscitation on the way to the hospital.

Do not allow any pressure-cycling mechanical resuscitator to be used with cardiac compression.

NOTES

15

OSHA § 1910.269

OSHA Regulations (Standards—29 CFR):
Electric Power Generation, Transmission,
and Distribution (1910.269)

- **Standard number:** 1910.269
- **Standard title:** Electric Power Generation, Transmission, and Distribution
- **Subpart:** R
- **Subpart title:** Special Industries

Interpretation(s)

(a) "General"

(a)(1) "Application"

(a)(1)(i) This section covers the operation and maintenance of electric power generation, control, transformation, transmission, and distribution lines and equipment. These provisions apply to:

(a)(1)(i)(A) Power generation, transmission, and distribution installations, including related equipment for the purpose of communication or metering, which are accessible only to qualified employees;
Note: The types of installations covered by this paragraph include the generation, transmission, and distribution installations of electric utilities, as well as equivalent installations of industrial establishments. Supplementary electric generating equipment that is used to supply a

workplace for emergency, standby, or similar purposes only is covered under Subpart S of this Part. (See paragraph (a)(1)(ii)(B) of this section.)

(a)(1)(i)(B) Other installations at an electric power generating station, as follows:

(a)(1)(i)(B)(1) Fuel and ash handling and processing installations, such as coal conveyors,

(a)(1)(i)(B)(2) Water and steam installations, such as penstocks, pipelines, and tanks, providing a source of energy for electric generators, and

(a)(1)(i)(B)(3) Chlorine and hydrogen systems:

(a)(1)(i)(C) Test sites where electrical testing involving temporary measurements associated with electric power generation, transmission, and distribution is performed in laboratories, in the field, in substations, and on lines, as opposed to metering, relaying, and routine line work;

(a)(1)(i)(D) Work on or directly associated with the installations covered in paragraphs (a)(1)(i)(A) through (a)(1)(i)(C) of this section; and

(a)(1)(i)(E) Line-clearance tree-trimming operations, as follows:

(a)(1)(i)(E)(1) Entire § 1910.269 of this Part, except paragraph (r)(1) of this section, applies to line-clearance tree-trimming operations performed by qualified employees (those who are knowledgeable in the construction and operation of electric power generation, transmission, or distribution equipment involved, along with the associated hazards).

(a)(1)(i)(E)(2) Paragraphs (a)(2), (b), (c), (g), (k), (p), and (r) of this section apply to line-clearance tree-trimming operations performed by line-clearance tree trimmers who are not qualified employees.

(a)(1)(ii) Notwithstanding paragraph (a)(1)(i) of this section, § 1910.269 of this Part does not apply:

(a)(1)(ii)(A) To construction work, as defined in § 1910.12 of this Part; or

(a)(1)(ii)(B) To electrical installations, electrical safety-related work practices, or electrical maintenance considerations covered by Subpart S of this Part.

Note 1: Work practices conforming to § 1910.332 through § 1910.335 of this Part are considered as complying with the electrical safety-related work practice requirements of this section identified in Table 1 of Appendix A.2 to this section, provided the work is being performed on a generation or distribution installation meeting § 1910.303 through § 1910.308 of this Part. This table also identifies provisions in this section that apply to work by qualified persons directly on or associated with installations of electric power generation, transmission, and distribution lines or equipment, regardless of compliance with § 1910.332 through § 1910.335 of this Part.

Note 2: Work practices performed by qualified persons and conforming to § 1910.269 of this Part are considered as complying with § 1910.333(c) and § 1910.335 of this Part.

(a)(1)(iii)   This section applies in addition to all other applicable standards contained in this Part § 1910. Specific references in this section to other sections of Part § 1910 are provided for emphasis only.

(a)(2)   "Training"

(a)(2)(i)   Employees shall be trained in and familiar with the safety-related work practices, safety procedures, and other safety requirements in this section that pertain to their respective job assignments. Employees shall also be trained in and familiar with any other safety practices, including applicable emergency procedures (such as pole-top and manhole rescue), that are not specifically addressed by this section but that are related to their work and are necessary for their safety.

(a)(2)(ii)   Qualified employees shall also be trained and competent in:

(a)(2)(ii)(A)   The skills and techniques necessary to distinguish exposed live parts from other parts of electric equipment.

(a)(2)(ii)(B)   The skills and techniques necessary to determine the nominal voltage of exposed live parts.

(a)(2)(ii)(C)   The minimum approach distances specified in this section corresponding to the voltages to which the qualified employee will be exposed, and

(a)(2)(ii)(D)   The proper use of the special precautionary techniques, personal protective equipment, insulating and shielding materials, and insulated tools for working on or near exposed energized parts of electric equipment.

Note: For the purposes of this section, a person must have this training in order to be considered a qualified person.

(a)(2)(iii)   The employer shall determine, through regular supervision and through inspections conducted on at least an annual basis, that each employee is complying with the safety-related work practices required by this section.

(a)(2)(iv)   An employee shall receive additional training (or retraining) under any of the following conditions:

(a)(2)(iv)(A)   If the supervision and annual inspections required by paragraph (a)(2)(iii) of this section indicate that the employee is not complying with the safety-related work practices required by this section, or

(a)(2)(iv)(B)   If new technology, new types of equipment, or changes in procedures necessitate the use of safety-related work practices that are different from those which the employee would normally use, or

(a)(2)(iv)(C)   If he or she must employ safety-related work practices that are not normally used during his or her regular job duties.

Note: OSHA would consider tasks that are performed less often than once per year to necessitate retraining before the performance of the work practices involved.

(a)(2)(v)   The training required by paragraph (a)(2) of this section shall be of the classroom or on-the-job type.

(a)(2)(vi)   The training shall establish employee proficiency in the work practices required by this section and shall introduce the procedures necessary for compliance with this section.

(a)(2)(vii)   The employer shall certify that each employee has received the training required by paragraph (a)(2) of this section. This certification shall be made when the employee demonstrates proficiency in the work practices involved and shall be maintained for the duration of the employee's employment.

Note: Employment records that indicate that an employee has received the required training are an acceptable means of meeting this requirement.

(a)(3)   "Existing conditions." Existing conditions related to the safety of the work to be performed shall be determined before work on or near electric lines or equipment is started. Such conditions include, but are not limited to, the nominal voltages of lines and equipment, the maximum switching transient voltages, the presence of hazardous induced voltages, the presence and condition of protective grounds and equipment-grounding conductors, the condition of poles, environmental conditions relative to safety, and the locations of circuits and equipment, including power and communication lines and fire protective signaling circuits.

(b) "Medical services and first aid." The employer shall provide medical services and first aid as required in § 1910.151 of this Part. In addition to the requirements of § 1910.151 of this Part, the following requirements also apply:

(b)(1) "Cardiopulmonary resuscitation and first-aid training." When employees are performing work on or associated with exposed lines or equipment energized at 50 volts or more, persons trained in first aid including cardiopulmonary resuscitation (CPR) shall be available as follows:

(b)(1)(i) For field work involving two or more employees at a work location, at least two trained persons shall be available. However, only one trained person need be available if all new employees are trained in first aid, including CPR, within 3 months of their hiring dates.

(b)(1)(ii) For fixed work locations such as generating stations, the number of trained persons available shall be sufficient to ensure that each employee exposed to electric shock can be reached within 4 minutes by a trained person. However, where the existing number of employees is insufficient to meet this requirement (at a remote substation, for example), all employees at the work location shall be trained.

(b)(2) "First-aid supplies." First-aid supplies required by § 1910.151(b) of this Part shall be placed in weatherproof containers if the supplies could be exposed to the weather.

(b)(3) "First-aid kits." Each first-aid kit shall be maintained, shall be readily available for use, and shall be inspected frequently enough to ensure that expended items are replaced but at least once per year.

(c) "Job briefing." The employer shall ensure that the employee in charge conducts a job briefing with the employees involved before they start each job. The briefing shall cover at least the following subjects: hazards associated with the job, work procedures involved, special precautions, energy source controls, and personal protective equipment requirements.

(c)(1) "Number of briefings." If the work or operations to be performed during the work day or shift are repetitive and similar, at least one job briefing shall be conducted before the start of the first job of each day or shift. Additional job briefings shall be held if significant changes, which might affect the safety of the employees, occur during the course of the work.

(c)(2) "Extent of briefing." A brief discussion is satisfactory if the work involved is routine and if the employee, by virtue of training and experience, can reasonably be expected to recognize and avoid the hazards involved in the job. A more extensive discussion shall be conducted:

(c)(2)(i) If the work is complicated or particularly hazardous, or

(c)(2)(ii) If the employee cannot be expected to recognize and avoid the hazards involved in the job.
   Note: The briefing is always required to touch on all the subjects listed in the introductory text to paragraph (c) of this section.

(c)(3) "Working alone." An employee working alone need not conduct a job briefing. However, the employer shall ensure that the tasks to be performed are planned as if a briefing were required.

(d) "Hazardous energy control (lockout/tagout) procedures"

(d)(1) "Application." The provisions of paragraph (d) of this section apply to the use of lockout/tagout procedures for the control of energy sources in installations for the purpose of electric power generation, including related equipment for communication or metering. Locking and tagging procedures for the de-energizing of electric energy sources, which are used exclusively for purposes of transmission and distribution are addressed by paragraph (m) of this section.
   Note 1: Installations in electric power generation facilities that are not an integral part of, or inextricably commingled with, power generation processes or equipment are covered under § 1910.147 and Subpart S of this Part.
   Note 2: Lockout and tagging procedures that comply with paragraphs (c) through (f) of § 1910.147 of this Part will also be deemed to comply with paragraph of this section if the procedures address the hazards covered by paragraph (d) of this section.

(d)(2) "General"

(d)(2)(i) The employer shall establish a program consisting of energy control procedures, employee training, and periodic inspections to ensure that, before any employee performs any servicing or maintenance on a machine or equipment where the unexpected energizing, start up, or release of stored energy could occur and cause injury, the machine or equipment is isolated from the energy source and rendered inoperative.

(d)(2)(ii)   The employer's energy-control program under paragraph (d)(2) of this section shall meet the following requirements:

(d)(2)(ii)(A)   If an energy-isolating device is not capable of being locked out, the employer's program shall use a tagout system.

(d)(2)(ii)(B)   If an energy-isolating device is capable of being locked out, the employer's program shall use lockout, unless the employer can demonstrate that the use of a tagout system will provide full employee protection as follows:

(d)(2)(ii)(B)(1)   When a tagout device is used on an energy-isolating device, which is capable of being locked out, the tagout device shall be attached at the same location that the lockout device would have been attached, and the employer shall demonstrate that the tagout program will provide a level of safety equivalent to that obtained by the use of a lockout program.

(d)(2)(ii)(B)(2)   In demonstrating that a level of safety is achieved in the tagout program equivalent to the level of safety obtained by the use of a lockout program, the employer shall demonstrate full compliance with all tagout-related provisions of this standard together with such additional elements as are necessary to provide the equivalent safety available from the use of a lockout device. Additional means to be considered as part of the demonstration of full employee protection shall include the implementation of additional safety measures such as the removal of an isolating circuit element, blocking of a controlling switch, opening of an extra disconnecting device, or the removal of a valve handle to reduce the likelihood of inadvertent energizing.

(d)(2)(ii)(C)   After November 1, 1994, whenever replacement or major repair, renovation, or modification of a machine or equipment is performed, and whenever new machines or equipment are installed, energy-isolating devices for such machines or equipment shall be designed to accept a lockout device.

(d)(2)(iii)   Procedures shall be developed, documented, and used for the control of potentially hazardous energy covered by paragraph (d) of this section.

(d)(2)(iv)   The procedure shall clearly and specifically outline the scope, purpose, responsibility, authorization, rules, and techniques to be applied to the control of hazardous energy, and the measures to enforce compliance including, but not limited to, the following:

(d)(2)(iv)(A)   A specific statement of the intended use of this procedure;

(d)(2)(iv)(B)   Specific procedural steps for shutting down, isolating, blocking, and securing machines or equipment to control hazardous energy;

(d)(2)(iv)(C)   Specific procedural steps for the placement, removal, and transfer of lockout devices or tagout devices and the responsibility for them; and

(d)(2)(iv)(D)   Specific requirements for testing a machine or equipment to determine and verify the effectiveness of lockout devices, tagout devices, and other energy-control measures.

(d)(2)(v)   The employer shall conduct a periodic inspection of the energy-control procedure at least annually to ensure that the procedure and the provisions of paragraph (d) of this section are being followed.

(d)(2)(v)(A)   The periodic inspection shall be performed by an authorized employee who is not using the energy-control procedure being inspected.

(d)(2)(v)(B)   The periodic inspection shall be designed to identify and correct any deviations or inadequacies.

(d)(2)(v)(C)   If lockout is used for energy control, the periodic inspection shall include a review, between the inspector and each authorized employee, of that employee's responsibilities under the energy-control procedure being inspected.

(d)(2)(v)(D)   Where tagout is used for energy control, the periodic inspection shall include a review, between the inspector and each authorized and affected employee, of that employee's responsibilities under the energy-control procedure being inspected, and the elements set forth in paragraph (d)(2)(vii) of this section.

(d)(2)(v)(E)   The employer shall certify that the inspections required by paragraph (d)(2)(v) of this section have been accomplished. The certification shall identify the machine or equipment on which the energy control procedure was being used, the date of the inspection, the employees included in the inspection, and the person performing the inspection.

Note: If normal work schedule and operation records demonstrate adequate inspection activity and contain the required information, no additional certification is required.

(d)(2)(vi) The employer shall provide training to ensure that the purpose and function of the energy-control program are understood by employees and that the knowledge and skills required for the safe application, usage, and removal of energy controls are acquired by employees. The training shall include the following:

(d)(2)(vi)(A) Each authorized employee shall receive training in the recognition of applicable hazardous energy sources, the type and magnitude of energy available in the workplace, and in the methods and means necessary for energy isolation and control.

(d)(2)(vi)(B) Each affected employee shall be instructed in the purpose and use of the energy control procedure.

(d)(2)(vi)(C) All other employees whose work operations are or may be in an area where energy-control procedures may be used shall be instructed about the procedures and about the prohibitions relating to attempts to restart or reenergize machines or equipments that are locked out or tagged out.

(d)(2)(vii) When tagout systems are used, employees shall also be trained in the following limitations of tags:

(d)(2)(vii)(A) Tags are essentially warning devices affixed to energy-isolating devices and do not provide the physical restraint on those devices that is provided by a lock.

(d)(2)(vii)(B) When a tag is attached to an energy-isolating means, it is not to be removed without authorization of the authorized person responsible for it, and it is never to be bypassed, ignored, or otherwise defeated.

(d)(2)(vii)(C) Tags must be legible and understandable by all authorized employees, affected employees, and all other employees whose work operations are or may be in the area, in order to be effective.

(d)(2)(vii)(D) Tags and their means of attachment must be made of materials which will withstand the environmental conditions encountered in the workplace.

(d)(2)(vii)(E) Tags may evoke a false sense of security, and their meaning needs to be understood as part of the overall energy-control program.

(d)(2)(vii)(F) Tags must be securely attached to energy-isolating devices so that they cannot be inadvertently or accidentally detached during use.

(d)(2)(viii) Retraining shall be provided by the employer as follows:

(d)(2)(viii)(A) Retraining shall be provided for all authorized and affected employees whenever there is a change in their job assignments, a change in machines, equipments, or processes that present a new hazard or whenever there is a change in the energy-control procedures.

(d)(2)(viii)(B) Retraining shall also be conducted whenever a periodic inspection under paragraph (d)(2)(v) of this section reveals, or whenever the employer has reason to believe, that there are deviations from or inadequacies in an employee's knowledge or use of the energy-control procedures.

(d)(2)(viii)(C) The retraining shall reestablish employee proficiency and shall introduce new or revised control methods and procedures, as necessary.

(d)(2)(ix) The employer shall certify that employee training has been accomplished and is being kept up-to-date. The certification shall contain each employee's name and dates of training.

(d)(3) "Protective materials and hardware"

(d)(3)(i) Locks, tags, chains, wedges, key blocks, adapter pins, self-locking fasteners, or other hardware shall be provided by the employer for isolating, securing, or blocking of machines or equipments from energy sources.

(d)(3)(ii) Lockout devices and tagout devices shall be singularly identified; shall be the only devices used for controlling energy; may not be used for other purposes; and shall meet the following requirements:

(d)(3)(ii)(A) Lockout devices and tagout devices shall be capable of withstanding the environment to which they are exposed for the maximum period of time that exposure is expected.

(d)(3)(ii)(A)(1) Tagout devices shall be constructed and printed so that exposure to weather conditions or wet and damp locations will not cause the tag to deteriorate or the message on the tag to become illegible.

(d)(3)(ii)(A)(2)   Tagout devices shall be so constructed as not to deteriorate when used in corrosive environments.

(d)(3)(ii)(B)   Lockout devices and tagout devices shall be standardized within the facility in at least one of the following criteria: color, shape, or size. Additionally, in the case of tagout devices, print and format shall be standardized.

(d)(3)(ii)(C)   Lockout devices shall be substantial enough to prevent removal without the use of excessive force or unusual techniques, such as with the use of bolt cutters or metal-cutting tools.

(d)(3)(ii)(D)   Tagout devices, including their means of attachment, shall be substantial enough to prevent inadvertent or accidental removal. Tagout device attachment means shall be of a non-reusable type, attachable by hand, self-locking, and non-releasable with a minimum unlocking strength of no less than 50 pounds and shall have the general design and basic characteristics of being at least equivalent to a one-piece, all-environment-tolerant nylon cable tie.

(d)(3)(ii)(E)   Each lockout device or tagout device shall include provisions for the identification of the employee applying the device.

(d)(3)(ii)(F)   Tagout devices shall warn against hazardous conditions if the machine or equipment is energized and shall include a legend such as the following: Do Not Start, Do Not Open, Do Not Close, Do Not Energize, Do Not Operate.

Note: For specific provisions covering accident prevention tags, see § 1910.145 of this Part.

(d)(4)   "Energy isolation." Lockout- and tagout-device application and removal may only be performed by the authorized employees who are performing the servicing or maintenance.

(d)(5)   "Notification." Affected employees shall be notified by the employer or authorized employee of the application and removal of lockout or tagout devices. Notification shall be given before the controls are applied and after they are removed from the machine or equipment.

Note: See also paragraph (d)(7) of this section, which requires that the second notification take place before the machine or equipment is reenergized.

(d)(6)   "Lockout/tagout application." The established procedures for the application of energy control (the lockout or tagout procedures) shall include the following elements and actions, and these procedures shall be performed in the following sequence:

(d)(6)(i) Before an authorized or affected employee turns off a machine or equipment, the authorized employee shall have knowledge of the type and magnitude of the energy, the hazards of the energy to be controlled, and the method or means to control the energy.

(d)(6)(ii) The machine or equipment shall be turned off or shut down using the procedures established for the machine or equipment. An orderly shut down shall be used to avoid any additional or increased hazards to employees as a result of the equipment stoppage.

(d)(6)(iii) All energy-isolating devices that are needed to control the energy to the machine or equipment shall be physically located and operated in such a manner as to isolate the machine or equipment from energy sources.

(d)(6)(iv) Lockout or tagout devices shall be affixed to each energy-isolating device by authorized employees.

(d)(6)(iv)(A) Lockout devices shall be attached in a manner that will hold the energy-isolating devices in a "safe" or "off" position.

(d)(6)(iv)(B) Tagout devices shall be affixed in such a manner as will clearly indicate that the operation or movement of energy-isolating devices from the "safe" or "off" position is prohibited.

(d)(6)(iv)(B)(1) Where tagout devices are used with energy-isolating devices designed with the capability of being locked out, the tag attachment shall be fastened at the same point at which the lock would have been attached.

(d)(6)(iv)(B)(2) Where a tag cannot be affixed directly to the energy-isolating device, the tag shall be located as close as safely possible to the device, in a position that will be immediately obvious to anyone attempting to operate the device.

(d)(6)(v) Following the application of lockout or tagout devices to energy-isolating devices, all potentially hazardous stored or residual energy shall be relieved, disconnected, restrained, or otherwise rendered safe.

(d)(6)(vi) If there is a possibility of reaccumulation of stored energy to a hazardous level, verification of isolation shall be continued until the servicing or maintenance is completed or until the possibility of such accumulation no longer exists.

(d)(6)(vii) Before starting work on machines or equipments that have been locked out or tagged out, the authorized employee shall verify that isolation and de-energizing of the machine or equipment have been accomplished. If normally energized parts will be exposed to contact by an employee while the machine or equipment is de-energized, a test shall be performed to ensure that these parts are de-energized.

(d)(7) "Release from lockout/tagout." Before lockout or tagout devices are removed and energy is restored to the machine or equipment, procedures shall be followed and actions taken by the authorized employees to ensure the following:

(d)(7)(i) The work area shall be inspected to ensure that nonessential items have been removed and that machine or equipment components are operationally intact.

(d)(7)(ii) The work area shall be checked to ensure that all employees have been safely positioned or removed.

(d)(7)(iii) After lockout or tagout devices have been removed and before a machine or equipment is started, affected employees shall be notified that the lockout or tagout devices have been removed.

(d)(7)(iv) Each lockout or tagout device shall be removed from each energy-isolating device by the authorized employee who applied the lockout or tagout device. However, if that employee is not available to remove it, the device may be removed under the direction of the employer, provided that specific procedures and training for such removal have been developed, documented, and incorporated into the employer's energy-control program. The employer shall demonstrate that the specific procedure provides a degree of safety equivalent to that provided by the removal of the device by the authorized employee who applied it. The specific procedure shall include at least the following elements:

(d)(7)(iv)(A) Verification by the employer that the authorized employee who applied the device is not at the facility;

(d)(7)(iv)(B) Making all reasonable efforts to contact the authorized employee to inform him or her that his or her lockout or tagout device has been removed; and

(d)(7)(iv)(C) Ensuring that the authorized employee has this knowledge before he or she resumes work at that facility.

(d)(8) "Additional requirements"

(d)(8)(i) If the lockout or tagout devices must be temporarily removed from energy-isolating devices and the machine or equipment must be energized to test or position the machine, equipment, or component thereof, the following sequence of actions shall be followed:

(d)(8)(i)(A) Clear the machine or equipment of tools and materials in accordance with paragraph (d)(7)(i) of this section;

(d)(8)(i)(B) Remove employees from the machine or equipment area in accordance with paragraphs (d)(7)(ii) and (d)(7)(iii) of this section;

(d)(8)(i)(C) Remove the lockout or tagout devices as specified in paragraph (d)(7)(iv) of this section;

(d)(8)(i)(D) Energize and proceed with the testing or positioning; and

(d)(8)(i)(E) De-energize all systems and reapply energy-control measures in accordance with paragraph (d)(6) of this section to continue the servicing or maintenance.

(d)(8)(ii) When servicing or maintenance is performed by a crew, craft, department, or other group, they shall use a procedure which affords the employees a level of protection equivalent to that provided by the implementation of a personal lockout or tagout device. Group lockout or tagout devices shall be used in accordance with the procedures required by paragraphs (d)(2)(iii) and (d)(2)(iv) of this section including, but not limited to, the following specific requirements:

(d)(8)(ii)(A) Primary responsibility shall be vested in an authorized employee for a set number of employees working under the protection of a group lockout or tagout device (such as an operations lock);

(d)(8)(ii)(B) Provision shall be made for the authorized employee to ascertain the exposure status of all individual group members with regard to the lockout or tagout of the machine or equipment;

(d)(8)(ii)(C) When more than one crew, craft, department, or other group is involved, assignment of overall job-associated lockout or tagout control responsibility shall be given to an authorized employee designated to coordinate affected work forces and ensure continuity of protection; and

(d)(8)(ii)(D)   Each authorized employee shall affix a personal lockout or tagout device to the group lockout device, group lockbox, or comparable mechanism when he or she begins work and shall remove those devices when he or she stops working on the machine or equipment being serviced or maintained.

(d)(8)(iii)   Procedures shall be used during shift or personnel changes to ensure the continuity of lockout or tagout protection, including provision for the orderly transfer of lockout or tagout device protection between off-going and on-coming employees, to minimize their exposure to hazards from the unexpected energizing or start-up of the machine or equipment or from the release of stored energy.

(d)(8)(iv)   Whenever outside servicing personnel are to be engaged in activities covered by paragraph (d) of this section, the on-site employer and the outside employer shall inform each other of their respective lockout or tagout procedures, and each employer shall ensure that his or her personnel understand and comply with restrictions and prohibitions of the energy-control procedures being used.

(d)(8)(v)   If energy-isolating devices are installed in a central location and are under the exclusive control of a system operator, the following requirements apply:

(d)(8)(v)(A)   The employer shall use a procedure that affords employees a level of protection equivalent to that provided by the implementation of a personal lockout or tagout device.

(d)(8)(v)(B)   The system operator shall place and remove lockout and tagout devices in place of the authorized employee under paragraphs (d)(4), (d)(6)(iv), and (d)(7)(iv) of this section.

(d)(8)(v)(C)   Provisions shall be made to identify the authorized employee who is responsible for (that is, being protected by) the lockout or tagout device, to transfer responsibility for lockout and tagout devices, and to ensure that an authorized employee requesting removal or transfer of a lockout or tagout device is the one responsible for it before the device is removed or transferred.

(e)   "Enclosed spaces." This paragraph covers enclosed spaces that may be entered by employees. It does not apply to vented vaults if a determination is made that the ventilation system is operating to protect employees before they enter the space. This paragraph applies to routine entry into enclosed spaces in lieu of the permit-space entry requirements contained in paragraphs (d) through (k) of § 1910.146 of this Part. If, after the precautions given in paragraphs (e) and (t) of this section are taken, the hazards remaining in the enclosed space endanger the life of an entrant or could interfere with escape from the space, then entry into the enclosed space shall meet the permit-space entry requirements of paragraphs (d) through (k) of § 1910.146 of this Part.

Note: Entries into enclosed spaces conducted in accordance with permit-space entry requirements of paragraphs (d) through (k) of § 1910.146 of this Part are considered as complying with paragraph (e) of this section.

(e)(1) "Safe work practices." The employer shall ensure the use of safe work practices for entry into and work in enclosed spaces and for rescue of employees from such spaces.

(e)(2) "Training." Employees who enter enclosed spaces or who serve as attendants shall be trained in the hazards of enclosed-space entry, in enclosed-space entry procedures, and in enclosed-space rescue procedures.

(e)(3) "Rescue equipment." Employers shall provide equipment to ensure the prompt and safe rescue of employees from the enclosed space.

(e)(4) "Evaluation of potential hazards." Before any entrance cover to an enclosed space is removed, the employer shall determine whether it is safe to do so by checking for the presence of any atmospheric pressure or temperature differences and by evaluating whether there might be a hazardous atmosphere in the space. Any conditions making it unsafe to remove the cover shall be eliminated before the cover is removed.

Note: The evaluation called for in this paragraph may take the form of a check of the conditions expected to be in the enclosed space. For example, the cover could be checked to see if it is hot and, if it is fastened in place, could be loosened gradually to release any residual pressure. A determination must also be made of whether conditions at the site could cause a hazardous atmosphere, such as an oxygen deficient or flammable atmosphere, to develop within the space.

(e)(5) "Removal of covers." When covers are removed from enclosed spaces, the opening shall be promptly guarded by a railing, temporary cover, or other barrier intended to prevent an accidental fall through the opening and to protect employees working in the space from objects entering the space.

(e)(6) "Hazardous atmosphere." Employees may not enter any enclosed space while it contains a hazardous atmosphere, unless the entry conforms to the generic permit-required confined spaces standard in § 1910.146 of this Part.

Note: The term "entry" is defined in § 1910.146(b) of this Part.

(e)(7) "Attendants." While work is being performed in the enclosed space, a person with first-aid training meeting paragraph (b) of this section shall be immediately available outside the enclosed space to render emergency assistance if there is reason to believe that a hazard may exist in the space or if a hazard exists because of traffic patterns in the area of the opening used for entry. That person is not precluded from performing other duties outside the enclosed space if these duties do not distract the attendant from monitoring employees within the space.

Note: See paragraph (t)(3) of this section for additional requirements on attendants for work in manholes.

(e)(8) "Calibration of test instruments." Test instruments used to monitor atmospheres in enclosed spaces shall be kept in calibration, with a minimum accuracy of + or − 10 percent.

(e)(9) "Testing for oxygen deficiency." Before an employee enters an enclosed space, the internal atmosphere shall be tested for oxygen deficiency with a direct-reading meter or similar instrument, capable of collection and immediate analysis of data samples without the need for off-site evaluation. If continuous forced air ventilation is provided, testing is not required provided that the procedures used ensure that employees are not exposed to the hazards posed by oxygen deficiency.

(e)(10) "Testing for flammable gases and vapors." Before an employee enters an enclosed space, the internal atmosphere shall be tested for flammable gases and vapors with a direct-reading meter or similar instrument capable of collection and immediate analysis of data samples without the need for off-site evaluation. This test shall be performed after the oxygen testing and ventilation required by paragraph (e)(9) of this section demonstrate that there is sufficient oxygen to ensure the accuracy of the test for flammability.

(e)(11) "Ventilation and monitoring." If flammable gases or vapors are detected or if an oxygen deficiency is found, forced air ventilation shall be used to maintain oxygen at a safe level and to prevent a hazardous concentration of flammable gases and vapors from accumulating. A continuous monitoring program to ensure that no increase in flammable gas or vapor concentration occurs may be followed in lieu of ventilation, if flammable gases or vapors are detected at safe levels.

Note: See the definition of hazardous atmosphere for guidance in determining whether or not a given concentration or a substance is considered to be hazardous.

(e)(12) "Specific ventilation requirements." If continuous forced air ventilation is used, it shall begin before entry is made and shall be maintained long enough to ensure that a safe atmosphere exists before employees are allowed to enter the work area. The forced air ventilation shall be so directed as to ventilate the immediate area where employees are present within the enclosed space and shall continue until all employees leave the enclosed space.

(e)(13) "Air supply." The air supply for the continuous forced air ventilation shall be from a clean source and may not increase the hazards in the enclosed space.

(e)(14) "Open flames." If open flames are used in enclosed spaces, a test for flammable gases and vapors shall be made immediately before the open flame device is used and at least once per hour while the device is used in the space. Testing shall be conducted more frequently if conditions present in the enclosed space indicate that once per hour is insufficient to detect hazardous accumulations of flammable gases or vapors.

Note: See the definition of hazardous atmosphere for guidance in determining whether or not a given concentration of a substance is considered to be hazardous.

(f) "Excavations." Excavation operations shall comply with Subpart P of Part § 1926 of this Chapter.

(g) "Personal protective equipment"

(g)(1) "General." Personal protective equipment shall meet the requirements of Subpart I of this Part.

(g)(2) "Fall protection"

(g)(2)(i) Personal fall-arrest equipment shall meet the requirements of Subpart M of Part § 1926 of this Chapter.

(g)(2)(ii) Body belts and safety straps for work positioning shall meet the requirements of § 1926.959 of this Chapter.

(g)(2)(iii) Body belts, safety straps, lanyards, lifelines, and body harnesses shall be inspected before use each day to determine that the equipment is in safe working condition. Defective equipment may not be used.

(g)(2)(iv) Lifelines shall be protected against being cut or abraded.

(g)(2)(v) Fall-arrest equipment, work-positioning equipment, or travel-restricting equipment shall be used by employees working at elevated locations more than 4 feet (1.2 m) above the ground on poles, towers, or similar structures if other fall protection has not been provided. Fall-protection equipment is not required to be used by a qualified employee climbing or changing location on poles, towers, or similar structures, unless conditions, such as, but not limited to, ice, high winds, the design of the structure (for example, no provision for holding on with hands), or the presence of contaminants on the structure, could cause the employee to lose his or her grip or footing.

Note 1: This paragraph applies to structures that support overhead electric power generation, transmission, and distribution lines and equipment. It does not apply to portions of buildings, such as loading docks, to electric equipment, such as transformers and capacitors, nor to aerial lifts. Requirements for fall protection associated with walking and working surfaces are contained in Subpart D of this Part; requirements for fall protection associated with aerial lifts are contained in § 1910.67 of this Part.

Note 2: Employees undergoing training are not considered "qualified employees" for the purposes of this provision. Unqualified employees (including trainees) are required to use fall protection any time they are more than 4 feet (1.2 m) above the ground.

(g)(2)(vi) The following requirements apply to personal fall-arrest systems:

(g)(2)(vi)(A) When stopping or arresting a fall, personal fall-arrest systems shall limit the maximum arresting force on an employee to 900 pounds (4 kN) if used with a body belt.

(g)(2)(vi)(B) When stopping or arresting a fall, personal fall-arrest systems shall limit the maximum arresting force on an employee to 1800 pounds (8 kN) if used with a body harness.

(g)(2)(vi)(C) Personal fall-arrest systems shall be rigged such that an employee can neither free fall more than 6 feet (1.8 m) nor contact any lower level.

(g)(2)(vii) If vertical lifelines or droplines are used, not more than one employee may be attached to any one lifeline.

(g)(2)(viii) Snaphooks may not be connected to loops made in webbing-type lanyards.

(g)(2)(ix) Snaphooks may not be connected to each other.

(h) "Ladders, platforms, step bolts, and manhole steps"

(h)(1) "General." Requirements for ladders contained in Subpart D of this Part apply, except as specifically noted in paragraph (h)(2) of this section.

(h)(2) "Special ladders and platforms." Portable ladders and platforms used on structures or conductors in conjunction with overhead line work need not meet paragraphs (d)(2)(i) and (d)(2)(iii) of § 1910.25 of this Part or paragraph (c)(3)(iii) of § 1910.26 of this Part. However, these ladders and platforms shall meet the following requirements:

(h)(2)(i) Ladders and platforms shall be secured to prevent their becoming accidentally dislodged.

(h)(2)(ii) Ladders and platforms may not be loaded in excess of the working loads for which they are designed.

(h)(2)(iii) Ladders and platforms may be used only in applications for which they were designed.

(h)(2)(iv) In the configurations in which they are used, ladders and platforms shall be capable of supporting without failure at least 2.5 times the maximum intended load.

(h)(3) "Conductive ladders." Portable metal ladders and other portable conductive ladders may not be used near exposed energized lines or equipment. However, in specialized high-voltage work, conductive ladders shall be used where the employer can demonstrate that nonconductive ladders would present a greater hazard than conductive ladders.

(i) "Hand and portable power tools"

(i)(1) "General." Paragraph (i)(2) of this section applies to electric equipment connected by cord and plug. Paragraph (i)(3) of this section applies to portable and vehicle-mounted generators used to supply cord- and plug-connected equipment. Paragraph (i)(4) of this section applies to hydraulic and pneumatic tools.

(i)(2)   "Cord- and plug-connected equipment"

(i)(2)(i)   Cord- and plug-connected equipment supplied by premises wiring is covered by Subpart S of this Part.

(i)(2)(ii)   Any cord- and plug-connected equipment supplied by other than premises wiring shall comply with one of the following in lieu of § 1910.243(a)(5) of this Part:

(i)(2)(ii)(A)   It shall be equipped with a cord containing an equipment-grounding conductor connected to the tool frame and to a means for grounding the other end (however, this option may not be used where the introduction of the ground into the work environment increases the hazard to an employee); or

(i)(2)(ii)(B)   It shall be of the double-insulated type conforming to Subpart S of this Part; or

(i)(2)(ii)(C)   It shall be connected to the power supply through an isolating transformer with an ungrounded secondary.

(i)(3)   "Portable and vehicle-mounted generators." Portable and vehicle-mounted generators used to supply cord- and plug-connected equipment shall meet the following requirements:

(i)(3)(i)   The generator may only supply equipment located on the generator or the vehicle and cord- and plug-connected equipment through receptables mounted on the generator or the vehicle.

(i)(3)(ii)   The noncurrent-carrying metal parts of equipment and the equipment-grounding conductor terminals of the receptacles shall be bonded to the generator frame.

(i)(3)(iii)   In the case of vehicle-mounted generators, the frame of the generator shall be bonded to the vehicle frame.

(i)(3)(iv)   Any neutral conductor shall be bonded to the generator frame.

(i)(4)   "Hydraulic and pneumatic tools"

(i)(4)(i)   Safe operating pressures for hydraulic and pneumatic tools, hoses, valves, pipes, filters, and fittings may not be exceeded.

Note: If any hazardous defects are present, no operating pressure would be safe, and the hydraulic or pneumatic equipment involved may not be used. In the absence of defects, the maximum rated operating pressure is the maximum safe pressure.

(i)(4)(ii)   A hydraulic or pneumatic tool used where it may contact exposed live parts shall be designed and maintained for such use.

(i)(4)(iii)   The hydraulic system supplying a hydraulic tool used where it may contact exposed live parts shall provide protection against loss of insulating value for the voltage involved due to the formation of a partial vacuum in the hydraulic line.

Note: Hydraulic lines without check valves having a separation of more than 35 feet (10.7 m) between the oil reservoir and the upper end of the hydraulic system promote the formation of a partial vacuum.

(i)(4)(iv)   A pneumatic tool used on energized electric lines or equipment or used where it may contact exposed live parts shall provide protection against the accumulation of moisture in the air supply.

(i)(4)(v)   Pressure shall be released before connections are broken, unless quick-acting, self-closing connectors are used. Hoses may not be kinked.

(i)(4)(vi)   Employees may not use any part of their bodies to locate or attempt to stop a hydraulic leak.

(j)   "Live-line tools"

(j)(1)   "Design of tools." Live-line tool rods, tubes, and poles shall be designed and constructed to withstand the following minimum tests:

(j)(1)(i)   100,000 volts per foot (3281 volts per centimeter) of length for 5 minutes if the tool is made of fiberglass-reinforced plastic (FRP), or

(j)(1)(ii)   75,000 volts per foot (2461 volts per centimeter) of length for 3 minutes if the tool is made of wood, or

(j)(1)(iii)   Other tests that the employer can demonstrate are equivalent.
Note: Live-line tools using rod and tube that meet ASTM F711-89, Standard Specification for Fiberglass-Reinforced Plastic (FRP) Rod and Tube Used in Live-Line Tools, conform to paragraph (j)(1)(i) of his section.

(j)(2)   "Condition of tools"

(j)(2)(i)   Each live-line tool shall be wiped clean and visually inspected for defects before use each day.

(j)(2)(ii)   If any defect or contamination that could adversely affect the insulating qualities or mechanical integrity of the live-line tool is present after wiping, the tool shall be removed from service and examined and tested according to paragraph (j)(2)(iii) of this section before being returned to service.

(j)(2)(iii)   Live-line tools used for primary employee protection shall be removed from service every 2 years and whenever required under paragraph (j)(2)(ii) of this section for examination, cleaning, repairing, and testing as follows:

(j)(2)(iii)(A)   Each tool shall be thoroughly examined for defects.

(j)(2)(iii)(B)   If a defect or contamination that could adversely affect the insulating qualities or mechanical integrity of the live-line tool is found, the tool shall be repaired and refinished or shall be permanently removed from service. If no such defects or contamination is found, the tool shall be cleaned and waxed.

(j)(2)(iii)(C)   The tool shall be tested in accordance with paragraphs (j)(2)(iii)(D) and (j)(2)(iii)(E) of this section under the following conditions:

(j)(2)(iii)(C)(1)   After the tool has been repaired or refinished; and

(j)(2)(iii)(C)(2)   After the examination if repair or refinishing is not performed, unless the tool is made of FRP rod or foam-filled FRP tube and the employer can demonstrate that the tool has no defects that could cause it to fail in use.

(j)(2)(iii)(D)   The test method used shall be designed to verify the tool's integrity along its entire working length and, if the tool is made of FRP, its integrity under wet conditions.

(j)(2)(iii)(E)   The voltage applied during the tests shall be as follows:

(j)(2)(iii)(E)(1)   75,000 volts per foot (2461 volts per centimeter) of length for 1 minute if the tool is made of fiberglass, or

(j)(2)(iii)(E)(2)   50,000 volts per foot (1640 volts per centimeter) of length for 1 minute if the tool is made of wood, or

(j)(3)   Other tests that the employer can demonstrate are equivalent.

Note: Guidelines for the examination, cleaning, repairing, and in-service testing of live-line tools are contained in the Institute of Electrical and Electronics Engineers Guide for In-Service Maintenance and Electrical Testing of Live-Line Tools, IEEE Std. 978-1984.

(k)   "Materials handling and storage"

(k)(1) "General." Material handling and storage shall conform to the requirements of Subpart N of this Part.

(k)(2)   "Materials storage near energized lines or equipment"

(k)(2)(i)   In areas not restricted to qualified persons only, materials or equipment may not be stored closer to energized lines or exposed energized parts of equipment than the following distances plus an amount providing for the maximum sag and side swing of all conductors and providing for the height and movement of material-handling equipment:

(k)(2)(i)(A)   For lines and equipment energized at 50 kV or less, the distance is 10 feet (305 cm).

(k)(2)(i)(B)   For lines and equipment energized at more than 50 kV, the distance is 10 feet (305 cm) plus 4 inches (10 cm) for every 10 kV over 50 kV.

(k)(2)(ii)   In areas restricted to qualified employees, material may not be stored within the working space about energized lines or equipment.

Note: Requirements for the size of the working space are contained in paragraphs (u)(1) and (v)(3) of this section.

(l)   "Working on or near exposed energized parts." This paragraph applies to work on exposed live parts, or near enough to them, to expose the employee to any hazard they present.

(l)(1)   "General." Only qualified employees may work on or with exposed energized lines or parts of equipment. Only qualified employees may work in areas containing unguarded, uninsulated energized lines or parts of equipment operating at 50 volts or more. Electric lines and equipment shall be considered and treated as energized unless the provisions of paragraph (d) or paragraph (m) of this section have been followed.

(l)(1)(i)   Except as provided in paragraph (l)(1)(ii) of this section, at least two employees shall be present while the following types of work are being performed:

(l)(1)(i)(A)   Installation, removal, or repair of lines that are energized at more than 600 volts,

(l)(1)(i)(B)   Installation, removal, or repair of de-energized lines if an employee is exposed to contact with other parts energized at more than 600 volts,

(l)(1)(i)(C)   Installation, removal, or repair of equipment, such as transformers, capacitors, and regulators, if an employee is exposed to contact with parts energized at more than 600 volts,

(l)(1)(i)(D)   Work involving the use of mechanical equipment, other than insulated aerial lifts, near parts energized at more than 600 volts, and

(l)(1)(i)(E)   Other work that exposes an employee to electrical hazards greater than or equal to those posed by operations that are specifically listed in paragraphs (l)(1)(i)(A) through (l)(1)(i)(D) of this section.

(l)(1)(ii)   Paragraph (l)(1)(i) of this section does not apply to the following operations:

(l)(1)(ii)(A)   Routine switching of circuits, if the employer can demonstrate that conditions at the site allow this work to be performed safely,

(l)(1)(ii)(B)   Work performed with live-line tools if the employee is positioned so that he or she is neither within reach of nor otherwise exposed to contact with energized parts, and

(l)(1)(ii)(C)   Emergency repairs to the extent necessary to safeguard the general public.

(l)(2)   "Minimum approach distances." The employer shall ensure that no employee approaches or takes any conductive object closer to exposed energized parts than set forth in Tables R.6 through R.10, unless:

(l)(2)(i)   The employee is insulated from the energized part (insulating gloves or insulating gloves and sleeves worn in accordance with paragraph (l)(3) of this section are considered insulation of the employee only with regard to the energized part upon which work is being performed), or

(l)(2)(ii)   The energized part is insulated from the employee and from any other conductive object at a different potential, or

TABLE R.6 AC Live-Line Work Minimum Approach Distance

Nominal voltage in kilovolts phase to phase	Distance			
	Phase to ground exposure		Phase to phase exposure	
	(ft–in)	(m)	(ft–in)	(m)
0.05 to 1.0	(4)	(4)	(4)	(4)
1.1 to 15.0	2–1	0.64	2–2	0.66
15.1 to 36.0	2–4	0.72	2–7	0.77
36.1 to 46.0	2–7	0.77	2–10	0.85
46.1 to 72.5	3–0	0.90	3–6	1.05
72.6 to 121	3–2	0.95	4–3	1.29
138 to 145	3–7	1.09	4–11	1.50
161 to 169	4–0	1.22	5–8	1.71
230 to 242	5–3	1.59	7–6	2.27
345 to 362	8–6	2.59	12–6	3.80
500 to 550	11–3	3.42	18–1	5.50
765 to 800	14–11	4.53	26–0	7.91

NOTE 1: These distances take into consideration the highest switching surge an employee will be exposed to on any system with air as the insulating medium and the maximum voltages shown.

NOTE 2: The clear live-line tool distance shall equal or exceed the values for the indicated voltage ranges.

NOTE 3: See App. B to this section for information on how the minimum approach distances listed in the tables were derived.

NOTE 4: Avoid contact.

TABLE R.7 AC Live-Line Work Minimum Approach Distance with Overvoltage Factor Phase-to-Ground Exposure

Maximum anticipated per-unit transient over voltage	Distance in feet–inches						
	Maximum phase-to-phase voltage in kilovolts						
	121	145	169	242	362	552	800
1.5	....	....	....	....	....	6–0	9–8
1.6	....	....	....	....	....	6–6	10–8
1.7	....	....	....	....	....	7–0	11–8
1.8	....	....	....	....	....	7–7	12–8
1.9	....	....	....	....	....	8–1	13–9
2.0	2–5	2–9	3–0	3–10	5–3	8–9	14–11
2.1	2–6	2–10	3–2	4–0	5–5	9–4	.....
2.2	2–7	2–11	3–3	4–1	5–9	9–11	.....
2.3	2–8	3–0	3–4	4–3	6–1	10–6	.....
2.4	2–9	3–1	3–5	4–5	6–4	11–3	.....
2.5	2–9	3–2	3–6	4–6	6–8	.....	.....
2.6	2–10	3–3	3–8	4–8	7–1	.....	.....
2.7	2–11	3–4	3–9	4–10	7–5	.....	.....
2.8	3–0	3–5	3–10	4–11	7–9	.....	.....
2.9	3–1	3–6	3–11	5–1	8–2	.....	.....
3.0	3–2	3–7	4–0	5–3	8–6	.....	.....

NOTE 1: The distance specified in this table may be applied only where the maximum anticipated per-unit transient overvoltage has been determined by engineering analysis and has been supplied by the employer. Table R.6 applies otherwise.

NOTE 2: The distances specified in the table are the air, bare-band, and live-line tool distances.

NOTE 3: See App. B to this section for information on how the minimum approach distances listed in the tables were derived and on how to calculate revised minimum approach distances based on the control of transient overvoltages.

TABLE R.8   AC Live-Line Work Minimum Approach Distance with Overvoltage Factor Phase-to-Phase Exposure

Maximum anticipated per-unit transient over voltage	Distance in feet–inches						
	Maximum phase-to-phase voltage in kilovolts						
	121	145	169	242	362	552	800
1.5	....	....	....	....	....	7–4	12–1
1.6	....	....	....	....	....	8–9	14–6
1.7	....	....	....	....	....	10–2	17–2
1.8	....	....	....	....	....	11–7	19–11
1.9	....	....	....	....	....	13–2	22–11
2.0	3–7	4–1	4–8	6–1	8–7	14–10	26–0
2.1	3–7	4–2	4–9	6–3	8–10	15–7	.....
2.2	3–8	4–3	4–10	6–4	9–2	16–4	.....
2.3	3–9	4–4	4–11	6–6	9–6	17–2	.....
2.4	3–10	4–5	5–0	6–7	9–11	18–1	.....
2.5	3–11	4–6	5–2	6–9	10–4	.....	.....
2.6	4–0	4–7	5–3	6–11	10–9	.....	.....
2.7	4–1	4–8	5–4	7–0	11–2	.....	.....
2.8	4–1	4–9	5–5	7–2	11–7	.....	.....
2.9	4–2	4–10	5–6	7–4	12–1	.....	.....
3.0	4–3	4–11	5–8	7–6	12–6	.....	.....

NOTE 1: The distance specified in this table may be applied only where the maximum anticipated per-unit transient overvoltage has been determined by engineering analysis and has been supplied by the employer. Table R.6 applies otherwise.
NOTE 2: The distances specified in this table are the air, bare-band, and live-line tool distances.
NOTE 3: See App. B to this section for information on how the minimum approach distances listed in the tables were derived and on how to calculate revised minimum approach distances based on the control of transient overvoltages.

TABLE R.9   DC Live-Line Work Minimum Approach Distance with Overvoltage Factor

Maximum anticipated per-unit transient overvoltage	Distance in feet–inches				
	Maximum line-to-ground voltage in kilovolts				
	250	400	500	600	750
1.5 or lower..........	3–8	5–3	6–9	8–7	11–10
1.6........................	3–10	5–7	7–4	9–5	13–1
1.7........................	4–1	6–0	7–11	10–3	14–4
1.8........................	4–3	6–5	8–7	11–2	15–9

NOTE 1: The distance specified in this table may be applied only where the maximum anticipated per-unit transient overvoltage has been determined by engineering analysis and has been supplied by the employer. However, if the transient overvoltage factor is not known, a factor of 1.8 shall be assumed.
NOTE 2: The distances specified in this table are the air, bare-hand, and live-line tool distances.

TABLE R.10   Altitude Correction Factor

| Altitude | | Correction | Altitude | | Correction |
ft	m	factor	ft	m	factor
3000	900	1.00	10000	3000	1.20
4000	1200	1.02	12000	3600	1.25
5000	1500	1.05	14000	4200	1.30
6000	1800	1.08	16000	4800	1.35
7000	2100	1.11	18000	5400	1.39
8000	2400	1.14	20000	6000	1.44
9000	2700	1.17			

NOTE: If the work is performed at elevations greater than 3000 ft (900 m) above mean sea level, the minimum approach distance shall be determined by multiplying the distances in Tables R.6 through R.9 by the correction factor corresponding to the altitude at which work is performed.

(l)(2)(iii)   The employee is insulated from any other exposed conductive object, as during live-line bare-hand work.

Note: Paragraphs (u)(5)(i) and (v)(5)(i) of this section contain requirements for the guarding and isolation of live parts. Parts of electric circuits that meet these two provisions are not considered as "exposed" unless a guard is removed or an employee enters the space intended to provide isolation from the live parts.

(l)(3)   "Type of insulation." If the employee is to be insulated from energized parts by the use of insulating gloves (under paragraph (l)(2)(i) of this section), insulating sleeves shall also be used. However, insulating sleeves need not be used under the following conditions:

(l)(3)(i)   If exposed energized parts on which work is not being performed are insulated from the employee; and

(l)(3)(ii)   If such insulation is placed from a position not exposing the employee's upper arm to contact with other energized parts.

(l)(4)   "Working position." The employer shall ensure that each employee, to the extent that other safety-related conditions at the work site permit, works in a position from which a slip or shock will not bring the employee's body into contact with exposed, uninsulated parts energized at a potential different from the employee.

(l)(5)   "Making connections." The employer shall ensure that connections are made as follows:

(I)(5)(i)   In connecting de-energized equipment or lines to an energized circuit by means of a conducting wire or device, an employee shall first attach the wire to the de-energized part;

(I)(5)(ii)   When disconnecting equipment or lines from an energized circuit by means of a conducting wire or device, an employee shall remove the source end first; and

(I)(5)(iii)   When lines or equipment are connected to or disconnected from energized circuits, loose conductors shall be kept away from exposed energized parts.

(I)(6)   "Apparel"

(I)(6)(i)   When work is performed within reaching distance of exposed energized parts of equipment, the employer shall ensure that each employee removes or renders non-conductive all exposed conductive articles, such as key or watch chains, rings, or wrist watches or bands, unless such articles do not increase the hazards associated with contact with the energized parts.

(I)(6)(ii)   The employer shall train each employee who is exposed to the hazards of flames or electric arcs in the hazards involved.

(I)(6)(iii)   The employer shall ensure that each employee who is exposed to the hazards of flames or electric arcs does not wear clothing that, when exposed to flames or electric arcs, could increase the extent of injury that would be sustained by the employee.
Note: Clothing made from the following types of fabrics, either alone or in blends, is prohibited by this paragraph, unless the employer can demonstrate that the fabric has been treated to withstand the conditions that may be encountered or that the clothing is worn in such a manner as to eliminate the hazard involved: acetate, nylon, polyester and rayon.

(I)(7)   "Fuse handling." When fuses must be installed or removed with one or both terminals energized at more than 300 volts or with exposed parts energized at more than 50 volts, the employer shall ensure that tools or gloves rated for the voltage are used. When expulsion-type fuses are installed with one or both terminals energized at more than 300 volts, the employer shall ensure that each employee wears eye protection meeting the requirements of Subpart I of this Part, uses a tool rated for the voltage, and is clear of the exhaust path of the fuse barrel.

(l)(8) "Covered (noninsulated) conductors." The requirements of this section which pertain to the hazards of exposed live parts also apply when work is performed in the proximity of covered (noninsulated) wires.

(l)(9) "Noncurrent-carrying metal parts." Noncurrent-carrying metal parts of equipment or devices, such as transformer cases and circuit-breaker housings, shall be treated as energized at the highest voltage to which they are exposed, unless the employer inspects the installation and determines that these parts are grounded before work is performed.

(l)(10) "Opening circuits under load." Devices used to open circuits under load conditions shall be designed to interrupt the current involved.

(m) "De-energizing lines and equipment for employee protection"

(m)(1) "Application." Paragraph (m) of this section applies to the de-energizing of transmission and distribution lines and equipment for the purpose of protecting employees. Control of hazardous energy sources used in the generation of electric energy is covered in paragraph (d) of this section. Conductors and parts of electric equipment that have been de-energized under procedures other than those required by paragraph (d) or (m) of this section, as applicable, shall be treated as energized.

(m)(2) "General"

(m)(2)(i) If a system operator is in charge of the lines or equipment and their means of disconnection, all of the requirements of paragraph (m)(3) of this section shall be observed, in the order given.

(m)(2)(ii) If no system operator is in charge of the lines or equipment and their means of disconnection, one employee in the crew shall be designated as being in charge of the clearance. All of the requirements of paragraph (m)(3) of this section apply, in the order given, except as provided in paragraph (m)(2)(iii) of this section. The employee in charge of the clearance shall take the place of the system operator, as necessary.

(m)(2)(iii) If only one crew will be working on the lines or equipment and if the means of disconnection is accessible and visible to and under the sole control of the employee in charge of the clearance, paragraphs (m)(3)(i), (m)(3)(iii), (m)(3)(iv), (m)(3)(viii), and (m)(3)(xii) of this section do not apply. Additionally, tags required by the remaining provisions of paragraph (m)(3) of this section need not be used.

(m)(2)(iv) Any disconnecting means that are accessible to persons outside the employer's control (for example, the general public) shall be rendered inoperable while they are open for the purpose of protecting employees.

(m)(3) "De-energizing lines and equipment"

(m)(3)(i) A designated employee shall make a request of the system operator to have the particular section of line or equipment de-energized. The designated employee becomes the employee in charge [as this term is used in paragraph (m)(3) of this section] and is responsible for the clearance.

(m)(3)(ii) All switches, disconnectors, jumpers, taps, and other means through which known sources of electric energy may be supplied to the particular lines and equipment to be de-energized shall be opened. Such means shall be rendered inoperable, unless its design does not so permit, and tagged to indicate that employees are at work.

(m)(3)(iii) Automatically and remotely controlled switches that could cause the opened disconnecting means to close shall also be tagged at the point of control. The automatic or remote-control feature shall be rendered inoperable, unless its design does not so permit.

(m)(3)(iv) Tags shall prohibit operation of the disconnecting means and shall indicate that employees are at work.

(m)(3)(v) After the applicable requirements in paragraphs (m)(3)(i) through (m)(3)(iv) of this section have been followed and the employee in charge of the work has been given a clearance by the system operator, the lines and equipment to be worked shall be tested to ensure that they are de-energized.

(m)(3)(vi) Protective grounds shall be installed as required by paragraph (n) of this section.

(m)(3)(vii) After the applicable requirements of paragraphs (m)(3)(i) through (m)(3)(vi) of this section have been followed, the lines and equipment involved may be worked as de-energized.

(m)(3)(viii) If two or more independent crews will be working on the same lines or equipment, each crew shall independently comply with the requirements in paragraph (m)(3) of this section.

(m)(3)(ix) To transfer the clearance, the employee in charge (or, if the employee in charge is forced to leave the work site due to illness or other emergency, the employee's supervisor) shall inform the system operator; employees in the crew shall be informed of the transfer; and the new employee in charge shall be responsible for the clearance.

(m)(3)(x)   To release a clearance, the employee in charge shall:

(m)(3)(x)(A)   Notify employees under his or her direction that the clearance is to be released;

(m)(3)(x)(B)   Determine that all employees in the crew are clear of the lines and equipment;

(m)(3)(x)(C)   Determine that all protective grounds installed by the crew have been removed; and

(m)(3)(x)(D)   Report this information to the system operator and release the clearance.

(m)(3)(xi)   The person releasing a clearance shall be the same person who requested the clearance, unless responsibility has been transferred under paragraph (m)(3)(ix) of this section.

(m)(3)(xii)   Tags may not be removed unless the associated clearance has been released under paragraph (m)(3)(x) of this section.

(m)(3)(xiii)   Only after all protective grounds have been removed, after all crews working on the lines or equipment have released their clearances, after all employees are clear of the lines and equipment, and after all protective tags have been removed from a given point of disconnection, may action be initiated to reenergize the lines or equipment at that point of disconnection.

(n)   "Grounding for the protection of employees"

(n)(1)   "Application." Paragraph (n) of this section applies to the grounding of transmission and distribution lines and equipment for the purpose of protecting employees. Paragraph (n)(4) of this section also applies to the protective grounding of other equipment as required elsewhere in this section.

(n)(2)   "General." For the employee to work-lines or equipment as de-energized, the lines or equipment shall be de-energized under the provisions of paragraph (m) of this section and shall be grounded as specified in paragraphs (n)(3) through (n)(9) of this section. However, if the employer can demonstrate that installation of a ground is impracticable or that the conditions resulting from the installation of a ground would present greater hazards than working without grounds, the lines and equipment may be treated as de-energized provided all of the following conditions are met:

(n)(2)(i) The lines and equipment have been de-energized under the provision of paragraph (m) of this section.

(n)(2)(ii) There is no possibility of contact with another energized source.

(n)(2)(iii) The hazard of induced voltage is not present.

(n)(3) "Equipotential Zone." Temporary protective grounds shall be placed at such locations and arranged in such a manner as to prevent each employee from being exposed to hazardous differences in electrical potential.

(n)(4) "Protective grounding equipment"

(n)(4)(i) Protective grounding equipment shall be capable of conducting the maximum fault current that could flow at the point of grounding for the time necessary to clear the fault. This equipment shall have an ampacity greater than or equal to that of No. 2 AWG copper.
Note: Guidelines for protective grounding equipment are contained in American Society for Testing and Materials Standard Specifications for Temporary Grounding Systems to Be Used on De-Energized Electric Power Lines and Equipment, ASTM F855-1990.

(n)(4)(ii) Protective grounds shall have an impedance low enough to cause immediate operation of protective devices in case of accidental energizing of the lines or equipment.

(n)(5) "Testing." Before any ground is installed, lines and equipment shall be tested and found absent of nominal voltage, unless a previously installed ground is present.

(n)(6) "Order of connection." When a ground is to be attached to a line or to an equipment, the ground-end connection shall be attached first, and then the other end shall be attached by means of a live-line tool.

(n)(7) "Order of removal." When a ground is to be removed, the grounding device shall be removed from the line or equipment using a live-line tool before the ground-end connection is removed.

(n)(8) "Additional precautions." When work is performed on a cable at a location remote from the cable terminal, the cable may not be grounded at the cable terminal if there is a possibility of hazardous transfer of potential should a fault occur.

(n)(9) "Removal of grounds for test." Grounds may be removed temporarily during tests. During the test procedure, the employer shall ensure that each employee uses insulating equipment and is isolated from any hazards involved, and the employer shall institute any additional measures as may be necessary to protect each exposed employee in case the previously grounded lines and equipment become energized.

(o) "Testing and test facilities"

(o)(1) "Application." Paragraph (o) of this section provides for safe work practices for high-voltage and high-power testing performed in laboratories, shops, and substations, and in the field and on electric transmission and distribution lines and equipment. It applies only to testing involving interim measurements utilizing high voltage, high power, or combinations of both, and not to testing involving continuous measurements as in routine metering, relaying, and normal line work.

Note: Routine inspection and maintenance measurements made by qualified employees are considered to be routine line work and are not included in the scope of paragraph (o) of this section, as long as the hazards related to the use of intrinsic high-voltage or high-power sources require only the normal precautions associated with routine operation and maintenance work required in the other paragraphs of this section. Two typical examples of such excluded test work procedures are "phasing-out" testing and testing for a "no-voltage" condition.

(o)(2) "General requirements"

(o)(2)(i) The employer shall establish and enforce work practices for the protection of each worker from the hazards of high-voltage or high-power testing at all test areas, temporary and permanent. Such work practices shall include, as a minimum, test area guarding, grounding, and the safe use of measuring and control circuits. A means providing for periodic safety checks of field test areas shall also be included. [See paragraph (o)(6) of this section.]

(o)(2)(ii) Employees shall be trained in safe work practices upon their initial assignment to the test area, with periodic reviews and updates provided as required by paragraph (a)(2) of this section.

(o)(3) "Guarding of test areas"

(o)(3)(i) Permanent test areas shall be guarded by walls, fences, or barriers designed to keep employees out of the test areas.

(o)(3)(ii) In field testing, or at a temporary test site where permanent fences and gates are not provided, one of the following means shall be used to prevent unauthorized employees from entering:

(o)(3)(ii)(A) The test area shall be guarded by the use of distinctively colored safety tape that is supported approximately waist high and to which safety signs are attached.

(o)(3)(ii)(B) The test area shall be guarded by a barrier or barricade that limits access to the test area to a degree equivalent, physically and visually, to the barricade specified in paragraph (o)(3)(ii)(A) of this section, or

(o)(3)(ii)(C) The test area shall be guarded by one or more test observers stationed so that the entire area can be monitored.

(o)(3)(iii) The barriers required by paragraph (o)(3)(ii) of this section shall be removed when the protection they provide is no longer needed.

(o)(3)(iv) Guarding shall be provided within test areas to control access to test equipment or to apparatus under test that may become energized as part of the testing by either direct or inductive coupling, in order to prevent accidental employee contact with energized parts.

(o)(4) "Grounding practices"

(o)(4)(i) The employer shall establish and implement safe grounding practices for the test facility.

(o)(4)(i)(A) All conductive parts accessible to the test operator during the time the equipment is operating at high voltage shall be maintained at ground potential except for portions of the equipment that are isolated from the test operator by guarding.

(o)(4)(i)(B) Wherever ungrounded terminals of test equipment or apparatus under test may be present, they shall be treated as energized until determined by tests to be de-energized.

(o)(4)(ii) Visible grounds shall be applied, either automatically or manually with properly insulated tools, to the high-voltage circuits after they are de-energized and before work is performed on the circuit or item or apparatus under test. Common ground connections shall be solidly connected to the test equipment and the apparatus under test.

(o)(4)(iii) In high-power testing, an isolated ground-return conductor system shall be provided so that no intentional passage of current, with its attendant voltage rise, can occur in the ground grid or in the earth. However, an isolated ground-return conductor need not be provided if the employer can demonstrate that both the following conditions are met:

(o)(4)(iii)(A) An isolated ground-return conductor cannot be provided due to the distance of the test site from the electric energy source, and

(o)(4)(iii)(B) Employees are protected from any hazardous step and touch potentials that may develop during the test.

Note: See App. C to this section for information on measures that can be taken to protect employees from hazardous step and touch potentials.

(o)(4)(iv) In tests in which grounding of test equipment by means of the equipment-grounding conductor located in the equipment power cord cannot be used due to increased hazards to test personnel or the prevention of satisfactory measurements, a ground that the employer can demonstrate affords equivalent safety shall be provided, and the safety ground shall be clearly indicated in the test set up.

(o)(4)(v) When the test area is entered after equipment is de-energized, a ground shall be placed on the high-voltage terminal and any other exposed terminals.

(o)(4)(v)(A) High capacitance equipment or apparatus shall be discharged through a resistor rated for the available energy.

(o)(4)(v)(B) A direct ground shall be applied to the exposed terminals when the stored energy drops to a level at which it is safe to do so.

(o)(4)(vi) If a test trailer or test vehicle is used in field testing, its chassis shall be grounded. Protection against hazardous touch potentials with respect to the vehicle, instrument panels, and other conductive parts accessible to employees shall be provided by bonding, insulation, or isolation.

(o)(5) "Control and measuring circuits"

(o)(5)(i) Control wiring, meter connections, test leads and cables may not be run from a test area unless they are contained in a grounded metallic sheath and terminated in a grounded metallic enclosure or unless other precautions are taken that the employer can demonstrate as ensuring equivalent safety.

(o)(5)(ii) Meters and other instruments with accessible terminals or parts shall be isolated from test personnel to protect against hazards arising from such terminals and parts becoming energized during testing. If this isolation is provided by locating test equipment in metal compartments with viewing windows, interlocks shall be provided to interrupt the power supply if the compartment cover is opened.

(o)(5)(iii) The routing and connections of temporary wiring shall be made secure against damages, accidental interruptions, and other hazards. To the maximum extent possible, signal, control, ground, and power cables shall be kept separate.

(o)(5)(iv) If employees will be present in the test area during testing, a test observer shall be present. The test observer shall be capable of implementing the immediate de-energizing of test circuits for safety purposes.

(o)(6) "Safety check"

(o)(6)(i) Safety practices governing employee work at temporary or field test areas shall provide for a routine check of such test areas for safety at the beginning of each series of tests.

(o)(6)(ii) The test operator in charge shall conduct these routine safety checks before each series of tests and shall verify at least the following conditions:

(o)(6)(ii)(A) That barriers and guards are in workable condition and are properly placed to isolate hazardous areas;

(o)(6)(ii)(B) That system test status signals, if used, are in operable condition;

(o)(6)(ii)(C) That test power disconnects are clearly marked and readily available in an emergency;

(o)(6)(ii)(D) That ground connections are clearly identifiable;

(o)(6)(ii)(E) That personal protective equipment is provided and used as required by Subpart I of this Part and by this section; and

(o)(6)(ii)(F) That signal, ground, and power cables are properly separated.

(p) "Mechanical equipment"

(p)(1) "General requirements"

(p)(1)(i) The critical safety components of mechanical elevating and rotating equipment shall receive a thorough visual inspection before use on each shift.
   Note: Critical safety components of mechanical elevating and rotating equipment are components whose failure would result in a free fall or free rotation of the boom.

(p)(1)(ii) No vehicular equipment having an obstructed view to the rear may be operated on off-highway job sites where any employee is exposed to the hazards created by the moving vehicle, unless:

(p)(1)(ii)(A) The vehicle has a reverse signal alarm audible above the surrounding noise level, or

(p)(1)(ii)(B) The vehicle is backed up only when a designated employee signals that it is safe to do so.

(p)(1)(iii) The operator of an electric line truck may not leave his or her position at the controls while a load is suspended, unless the employer can demonstrate that no employee (including the operator) might be endangered.

(p)(1)(iv) Rubber-tired, self-propelled scrapers, rubber-tired front-end loaders, rubber-tired dozers, wheel-type agricultural and industrial tractors, crawler-type tractors, crawler-type loaders, and motor graders, with or without attachments, shall have roll-over protective structures that meet the requirements of Subpart W of Part § 1926 of this Chapter.

(p)(2) "Outriggers"

(p)(2)(i) Vehiclular equipment, if provided with outriggers, shall be operated with the outriggers extended and firmly set as necessary for the stability of the specific configuration of the equipment. Outriggers may not be extended or retracted outside of clear view of the operator unless all employees are outside the range of possible equipment motion.

(p)(2)(ii)   If the work area or the terrain precludes the use of outriggers, the equipment may be operated only within its maximum load ratings for the particular configuration of the equipment without outriggers.

(p)(3)   "Applied loads." Mechanical equipment used to lift or move lines or other material shall be used within its maximum load rating and other design limitations for the conditions under which the work is being performed.

(p)(4)   "Operations near energized lines or equipment"

(p)(4)(i)   Mechanical equipment shall be operated so that the minimum approach distances of Tables R.6 through R.10 are maintained for exposed energized lines and equipment. However, the insulated portion of an aerial lift operated by a qualified employee in the lift is exempted from this requirement.

(p)(4)(ii)   A designated employee other than the equipment operator shall observe the approach distance to exposed lines and equipment and give timely warnings before the minimum approach distance required by paragraph (p)(4)(i) is reached, unless the employer can demonstrate that the operator can accurately determine that the minimum approach distance is being maintained.

(p)(4)(iii)   If, during operation of the mechanical equipment, the equipment could become energized, the operation shall also comply with at least one of paragraphs (p)(4)(iii)(A) through (p)(4)(iii)(C) of this section.

(p)(4)(iii)(A)   The energized lines exposed to contact shall be covered with insulating protective material that will withstand the type of contact that might be made during the operation.

(p)(4)(iii)(B)   The equipment shall be insulated for the voltage involved. The equipment shall be positioned so that its uninsulated portions cannot approach the lines or equipment any closer than the minimum approach distances specified in Tables R.6 through R.10.

(p)(4)(iii)(C)   Each employee shall be protected from hazards that might arise from equipment contact with the energized lines. The measures used shall ensure that employees will not be exposed to hazardous differences in potential. Unless the employer can demonstrate that the methods in use protect each employee from the hazards that might arise if the equipment contacts the energized line, the measures used shall include all of the following techniques:

(p)(4)(iii)(C)(1) Using the best available ground to minimize the time the lines remain energized,

(p)(4)(iii)(C)(2) Bonding equipment together to minimize potential differences,

(p)(4)(iii)(C)(3) Providing ground mats to extend areas of equipotential, and

(p)(4)(iii)(C)(4) Employing insulating protective equipment or barricades to guard against any remaining hazardous potential differences.

Note: Appendix C to this section contains information on hazardous step and touch potentials and on methods of protecting employees from hazards resulting from such potentials.

(q) "Overhead lines." This paragraph provides additional requirements for work performed on or near overhead lines and equipment.

(q)(1) "General"

(q)(1)(i) Before elevated structures, such as poles or towers, are subjected to such stresses as climbing or the installation or removal of equipment may impose, the employer shall ascertain that the structures are capable of sustaining the additional or unbalanced stresses. If the pole or other structure cannot withstand the loads which will be imposed, it shall be braced or otherwise supported so as to prevent failure.

Note: Appendix D to this section contains test methods that can be used in ascertaining whether a wood pole is capable of sustaining the forces that would be imposed by an employee climbing the pole. This paragraph also requires the employer to ascertain that the pole can sustain all other forces that will be imposed by the work to be performed.

(q)(1)(ii) When poles are set, moved, or removed near exposed energized overhead conductors, the pole may not contact the conductors.

(q)(1)(iii) When a pole is set, moved, or removed near an exposed energized overhead conductor, the employer shall ensure that each employee wears electrical protective equipment or uses insulated devices when handling the pole and that no employee contacts the pole with uninsulated parts of his or her body.

(q)(1)(iv) To protect employees from falling into holes into which poles are to be placed, the holes shall be attended by employees or physically guarded whenever anyone is working nearby.

(q)(2) "Installing and removing overhead lines." The following provisions apply to the installation and removal of overhead conductors or cables:

(q)(2)(i) The employer shall use the tension stringing method, barriers, or other equivalent measures to minimize the possibility that conductors and cables being installed or removed will contact energized power lines or equipment.

(q)(2)(ii) The protective measures required by paragraph (p)(4)(iii) of this section for mechanical equipment shall also be provided for conductors, cables, and pulling and tensioning equipment when the conductor or cable is being installed or removed close enough to energized conductors that any of the following failures could energize the pulling or tensioning equipment or the wire or cable being installed or removed:

(q)(2)(ii)(A) Failure of the pulling or tensioning equipment,

(q)(2)(ii)(B) Failure of the wire or cable being pulled, or

(q)(2)(ii)(C) Failure of the previously installed lines or equipment.

(q)(2)(iii) If the conductors being installed or removed crossover energized conductors in excess of 600 volts and if the design of the circuit-interrupting devices protecting the lines so permits, the automatic-reclosing feature of these devices shall be made inoperative.

(q)(2)(iv) Before lines are installed parallel to existing energized lines, the employer shall make a determination of the approximate voltage to be induced in the new lines, or work shall proceed on the assumption that the induced voltage is hazardous. Unless the employer can demonstrate that the lines being installed are not subject to the induction of a hazardous voltage or unless the lines are treated as energized, the following requirements also apply:

(q)(2(iv)(A) Each bare conductor shall be grounded in increments so that no point along the conductor is more than 2 miles (3.22 km) from a ground.

(q)(2)(iv)(B) The grounds required in paragraph (q)(2)(iv)(A) of this section shall be left in place until the conductor installation is completed between dead ends.

(q)(2)(iv)(C) The grounds required in paragraph (q)(2)(iv)(A) of this section shall be removed as the last phase of aerial cleanup.

(q)(2)(iv)(D)   If employees are working on bare conductors, grounds shall also be installed at each location where these employees are working, and grounds shall be installed at all open dead-end or catch-off points or the next adjacent structure.

(q)(2)(iv)(E)   If two bare conductors are to be spliced, the conductors shall be bonded and grounded before being spliced.

(q)(2)(v)   Reel-handling equipment, including pulling and tensioning devices, shall be in safe operating condition and shall be leveled and aligned.

(q)(2)(vi)   Load ratings of stringing lines, pulling lines, conductor grips, load-bearing hardware and accessories, rigging, and hoists may not be exceeded.

(q)(2)(vii)   Pulling lines and accessories shall be repaired or replaced when defective.

(q)(2)(viii)   Conductor grips may not be used on wire rope, unless the grip is specifically designed for this application.

(q)(2)(ix)   Reliable communications, through two-way radios or other equivalent means, shall be maintained between the reel tender and the pulling rig operator.

(q)(2)(x)   The pulling rig may only be operated when it is safe to do so. *Note:* Examples of unsafe conditions include employees in locations prohibited by paragraph (q)(2)(xi) of this section, conductor and pulling line hang-ups, and slipping of the conductor grip.

(q)(2)(xi)   While the conductor or pulling line is being pulled (in motion) with a power-driven device, employees are not permitted directly under overhead operations or on the cross arm, except as necessary to guide the stringing sock or board over or through the stringing sheave.

(q)(3)   "Live-line bare-hand work." In addition to other applicable provisions contained in this section, the following requirements apply to live-line bare-hand work:

(q)(3)(i)   Before using or supervising the use of the live-line bare-hand technique on energized circuits, employees shall be trained in the technique and in the safety requirements of paragraph (q)(3) of this section. Employees shall receive refresher training as required by paragraph (a)(2) of this section.

(q)(3)(ii)   Before any employee uses the live-line bare-hand technique on energized high-voltage conductors or parts, the following information shall be ascertained:

(q)(3)(ii)(A)   The nominal voltage rating of the circuit on which the work is to be performed.

(q)(3)(ii)(B)   The minimum approach distances to ground of lines and other energized parts on which work is to be performed, and

(q)(3)(ii)(C)   The voltage limitations of equipment to be used.

(q)(3)(iii)   The insulated equipment, insulated tools, and aerial devices and platforms used shall be designed, tested, and intended for live-line bare-hand work. Tools and equipment shall be kept clean and dry while they are in use.

(q)(3)(iv)   The automatic-reclosing feature of circuit-interrupting devices protecting the lines shall be made inoperative, if the design of the devices permits.

(q)(3)(v)   Work may not be performed when adverse weather conditions would make the work hazardous even after the work practices required by this section are employed. Additionally, work may not be performed when winds reduce the phase-to-phase or phase-to-ground minimum approach distances at the work location below that specified in paragraph (q)(3)(xiii) of this section, unless the grounded objects and other lines and equipment are covered by insulating guards.
   Note: Thunderstorms in the immediate vicinity, high winds, snow storms, and ice storms are examples of adverse weather conditions that are presumed to make live-line bare-hand work too hazardous to perform safely.

(q)(3)(vi)   A conductive bucket liner or other conductive device shall be provided for bonding the insulated aerial device to the energized line or equipment.

(q)(3)(vi)(A)   The employee shall be connected to the bucket liner or other conductive device by the use of conductive shoes, leg clips, or other means.

(q)(3)(vi)(B)   Where differences in potentials at the work site pose a hazard to employees, electrostatic shielding designed for the voltage being worked shall be provided.

(q)(3)(vii)   Before the employee contacts the energized part, the conductive bucket liner or other conductive device shall be bonded to the energized conductor by means of a positive connection. This connection shall remain attached to the energized conductor until the work on the energized circuit is completed.

(q)(3)(viii)   Aerial lifts to be used for live-line bare-hand work shall have dual controls (lower and upper) as follows:

(q)(3)(viii)(A)   The upper controls shall be within easy reach of the employee in the bucket. On a two-bucket-type lift, across to the controls shall be within easy reach from either bucket.

(q)(3)(viii)(B)   The lower set of controls shall be located near the base of the boom, and they shall be so designed that they can override operation of the equipment at any time.

(q)(3)(ix)   Lower (ground-level) lift controls may not be operated with an employee in the lift, except in case of emergency.

(q)(3)(x)   Before employees are elevated into the work position, all controls (ground level and bucket) shall be checked to determine that they are in proper working condition.

(q)(3)(xi)   Before the boom of an aerial lift is elevated, the body of the truck shall be grounded, or the body of the truck shall be barricaded and treated as energized.

(q)(3)(xii)   A boom-current test shall be made before work is started each day, each time during the day when higher voltage is encountered, and when changed conditions indicate a need for an additional test. This test shall consist of placing the bucket in contact with an energized source equal to the voltage to be encountered for a minimum of 3 minutes. The leakage current may not exceed 1 microampere per kilovolt of nominal phase-to-ground voltage. Work from the aerial lift shall be immediately suspended upon indication of a malfunction in the equipment.

(q)(3)(xiii)   The minimum approach distances specified in Tables R.6 through R.10 shall be maintained from all grounded objects and from lines and equipment at a potential different from that to which the live-line bare-hand equipment is bonded, unless such grounded objects and other lines and equipment are covered by insulating guards.

(q)(3)(xiv)   While an employee is approaching, leaving, or bonding to an energized circuit, the minimum approach distances in Tables R.6 through R.10 shall be maintained between the employee and any grounded parts, including the lower boom and portions of the truck.

(q)(3)(xv)   While the bucket is positioned alongside an energized bushing or insulator string, the phase-to-ground minimum approach distances to Tables R.6 through R.10 shall be maintained between all parts of the bucket and the grounded end of the bushing or insulator string or any other grounded surface.

(q)(3)(xvi)   Hand lines may not be used between the bucket and the boom or between the bucket and the ground. However, non-conductive-type hand lines may be used from conductor to ground if not supported from the bucket. Ropes used for live-line bare-hand work may not be used for other purposes.

(q)(3)(xvii)   Uninsulated equipment or material may not be passed between a pole or structure and an aerial lift while an employee working from the bucket is bonded to an energized part.

(q)(3)(xviii)   A minimum approach distance table reflecting the minimum approach distances listed in Tables R.6 through R.10 shall be printed on a plate of durable non-conductive material. This table shall be mounted so as to be visible to the operator of the boom.

(q)(3)(xix)   A non-conductive measuring device shall be readily accessible to assist employees in maintaining the required minimum approach distance.

(q)(4)   "Towers and structures." The following requirements apply to work performed on towers or other structures which support overhead lines:

(q)(4)(i)   The employer shall ensure that no employee is under a tower or structure while work is in progress, except where the employer can demonstrate that such a working position is necessary to assist employees working above.

(q)(4)(ii)   Tag lines or other similar devices shall be used to maintain control of tower sections being raised or positioned, unless the employer can demonstrate that the use of such devices would create a greater hazard.

(q)(4)(iii)   The load line may not be detached from a member or section until the load is safely secured.

(q)(4)(iv)   Except during emergency restoration procedures, work shall be discontinued when adverse weather conditions would make the work hazardous in spite of the work practices required by this section.

Note: Thunderstorms in the immediate vicinity, high winds, snow storms, and ice storms are examples of adverse weather conditions that are presumed to make this work too hazardous to perform, except under emergency conditions.

(r)   "Line-clearance tree-trimming operations." This paragraph provides additional requirements for line-clearance tree-trimming operations and for equipment used in these operations.

(r)(1)   "Electrical hazards." This paragraph does not apply to qualified employees.

(r)(1)(i)   Before an employee climbs, enters, or works around any tree, a determination shall be made of the nominal voltage of electric power lines posing a hazard to employees. However, a determination of the maximum nominal voltage to which an employee will be exposed may be made instead, if all lines are considered as energized at this maximum voltage.

(r)(1)(ii)   There shall be a second line-clearance tree trimmer within normal (that is, unassisted) voice communication under any of the following conditions:

(r)(1)(ii)(A)   If a line-clearance tree trimmer is to approach more closely than 10 feet (305 cm) any conductor or electric apparatus energized at more than 750 volts, or

(r)(1)(ii)(B)   If branches or limbs being removed are closer to lines energized at more than 750 volts than the distances listed in Tables R.6, R.9, and R.10, or

(r)(1)(ii)(C)   If roping is necessary to remove branches or limbs from such conductors or apparatus.

(r)(1)(iii)   Line-clearance tree trimmers shall maintain the minimum approach distances from energized conductors given in Tables R.6, R.9, and R.10.

(r)(1)(iv)   Branches that are contacting exposed energized conductors or equipment or that are within the distances specified in Tables R.6,

R.9, and R.10 may be removed only through the use of insulating equipment.

Note: A tool constructed of a material that the employer can demonstrate has insulating qualities meeting paragraph (j)(1) of this section is considered as insulated under this paragraph if the tool is clean and dry.

(r)(1)(v) Ladders, platforms, and aerial devices may not be brought closer to an energized part than the distances listed in Tables R.6, R.9, and R.10.

(r)(1)(vi) Line-clearance tree-trimming work may not be performed when adverse weather conditions make the work hazardous in spite of the work practices required by this section. Each employee performing line-clearance tree-trimming work in the aftermath of a storm or under similar emergency conditions shall be trained in the special hazards related to this type of work.

Note: Thunderstorms in the immediate vicinity, high winds, snow storms, and ice storms are examples of adverse weather conditions that are presumed to make line-clearance tree-trimming work too hazardous to perform safely.

(r)(2) "Brush chippers"

(r)(2)(i) Brush chippers shall be equipped with a locking device in the ignition system.

(r)(2)(ii) Access panels for maintenance and adjustment of the chipper blades and associated drive train shall be in place and secure during operation of the equipment.

(r)(2)(iii) Brush chippers not equipped with a mechanical infeed system shall be equipped with an infeed hopper of length sufficient to prevent employees from contacting the blades or knives of the machine during operation.

(r)(2)(iv) Trailer chippers detached from trucks shall be chocked or otherwise secured.

(r)(2)(v) Each employee in the immediate area of an operating chipper feed table shall wear personal protective equipment as required by Subpart I of this Part.

(r)(3) "Sprayers and related equipment"

(r)(3)(i) Walking and working surfaces of sprayers and related equipment shall be covered with slip-resistant material. If slipping hazards cannot be eliminated, slip-resistant footwear or handrails and stair rails meeting the requirements of Subpart D may be used instead of slip-resistant material.

(r)(3)(ii) Equipment on which employees stand to spray while the vehicle is in motion shall be equipped with guardrails around the working area. The guardrail shall be constructed in accordance with Subpart D of this Part.

(r)(4) "Stump cutters"

(r)(4)(i) Stump cutters shall be equipped with enclosures or guards to protect employees.

(r)(4)(ii) Each employee in the immediate area of stump grinding operations (including the stump cutter operator) shall wear personal protective equipment as required by Subpart I of this Part.

(r)(5) "Gasoline-engine power saws." Gasoline-engine power-saw operations shall meet the requirements of § 1910.266(c)(5) of this part and the following:

(r)(5)(i) Each power saw weighing more than 15 pounds (6.8 kilograms, service weight) that is used on trees shall be supported by a separate line, except when work is performed from an aerial lift and except during topping or removing operations where no supporting limb will be available.

(r)(5)(ii) Each power saw shall be equipped with a control that will return the saw to idling speed when released.

(r)(5)(iii) Each power saw shall be equipped with a clutch and shall be so adjusted that the clutch will not engage the chain drive at idling speed.

(r)(5)(iv) A power saw shall be started on the ground or where it is otherwise firmly supported. Drop starting of saws over 15 pounds (6.8 kg) is permitted outside of the bucket of an aerial lift only if the area below the lift is clear of personnel.

(r)(5)(v) A power saw engine may be started and operated only when all employees other than the operator are clear of the saw.

(r)(5)(vi) A power saw may not be running when the saw is being carried up into a tree by an employee.

(r)(5)(vii) Power saw engines shall be stopped for all cleaning, refueling, adjustments, and repairs to the saw or motor, except as the manufacturer's servicing procedures require otherwise.

(r)(6) "Backpack power units for use in pruning and clearing"

(r)(6)(i) While a backpack power unit is running, no one other than the operator may be within 10 feet (305 cm) of the cutting head of a brush saw.

(r)(6)(ii) A backpack power unit shall be equipped with a quick shutoff switch readily accessible to the operator.

(r)(6)(iii) Backpack power unit engines shall be stopped for all cleaning, refueling, adjustments, and repairs to the saw or motor, except as the manufacturer's servicing procedures require otherwise.

(r)(7) "Rope"

(r)(7)(i) Climbing ropes shall be used by employees working aloft in trees. These ropes shall have a minimum diameter of 0.5 inch (1.2 cm) with a minimum breaking strength of 2300 pounds (10.2 kN). Synthetic rope shall have elasticity of not more than 7 percent.

(r)(7)(ii) Rope shall be inspected before each use and, if unsafe (for example, because of damage or defect), may not be used.

(r)(7)(iii) Rope shall be stored away from cutting edges and sharp tools. Rope contact with corrosive chemicals, gas, and oil shall be avoided.

(r)(7)(iv) When stored, rope shall be coiled and piled, or shall be suspended, so that air can circulate through the coils.

(r)(7)(v) Rope ends shall be secured to prevent their unraveling.

(r)(7)(vi) Climbing rope may not be spliced to effect repair.

(r)(7)(vii) A rope that is wet, that is contaminated to the extent that its insulating capacity is impaired, or that is otherwise not considered to be insulated for the voltage involved may not be used near exposed energized lines.

(r)(8) "Fall protection." Each employee shall be tied in with a climbing rope and safety saddle when the employee is working above the ground in a tree, unless he or she is ascending into the tree.

(s) "Communication facilities"

(s)(1) "Microwave transmission"

(s)(1)(i) The employer shall ensure that no employee looks into an open waveguide or antenna that is connected to an energized microwave source.

(s)(1)(ii) If the electromagnetic radiation level within an accessible area associated with microwave communications systems exceeds the radiation protection guide given in § 1910.97(a)(2) of this Part, the area shall be posted with the warning symbol described in § 1910.97(a)(3) of this Part. The lower half of the warning symbol shall include the following statements or ones that the employer can demonstrate are equivalent:
Radiation in this area may exceed hazard limitations and special precautions are required. Obtain specific instruction before entering.

(s)(1)(iii) When an employee works in an area where the electromagnetic radiation could exceed the radiation-protection guide, the employer shall institute measures that ensure that the employee's exposure is not greater than that permitted by that guide. Such measures may include administrative and engineering controls and personal protective equipment.

(s)(2) "Power line carrier." Power line carrier work, including work on equipment used for coupling carrier current to power line conductors, shall be performed in accordance with the requirements of this section pertaining to work on energized lines.

(t) "Underground electrical installations." This paragraph provides additional requirements for work on underground electrical installations.

(t)(1) "Access." A ladder or other climbing device shall be used to enter and exit a manhole or subsurface vault exceeding 4 feet (122 cm) in depth. No employee may climb into or out of a manhole or vault by stepping on cables or hangers.

(t)(2) "Lowering equipment into manholes." Equipment used to lower materials and tools into manholes or vaults shall be capable of supporting the weight to be lowered and shall be checked for defects before use.

Before tools or material are lowered into the opening for a manhole or vault, each employee working in the manhole or vault shall be clear of the area directly under the opening.

(t)(3) "Attendants for manholes"

(t)(3)(i) While work is being performed in a manhole containing energized electric equipment, an employee with first-aid and CPR training meeting paragraph (b)(1) of this section shall be available on the surface in the immediate vicinity to render emergency assistance.

(t)(3)(ii) Occasionally, the employee on the surface may briefly enter a manhole to provide assistance, other than emergency.

Note 1: An attendant may also be required under paragraph (e)(7) of this section. One person may serve to fulfill both requirements. However, attendants required under paragraph (e)(7) of this section are not permitted to enter the manhole.

Note 2: Employees entering manholes containing unguarded, uninsulated energized lines or parts of electric equipment operating at 50 volts or more are required to be qualified under paragraph (l)(1) of this section.

(t)(3)(iii) For the purpose of inspection, housekeeping, taking readings, or similar work, an employee working alone may enter, for brief periods of time, a manhole where energized cables or equipments are in service, if the employer can demonstrate that the employee will be protected from all electrical hazards.

(t)(3)(iv) Reliable communications, through two-way radios or other equivalent means, shall be maintained among all employees involved in the job.

(t)(4) "Duct rods." If duct rods are used, they shall be installed in the direction presenting the least hazard to employees. An employee shall be stationed at the far end of the duct line being rodded to ensure that the required minimum approach distances are maintained.

(t)(5) "Multiple cables." When multiple cables are present in a work area, the cable to be worked shall be identified by electrical means, unless its identity is obvious by reason of distinctive appearance or location or by other readily apparent means of identification. Cables other than the one being worked shall be protected from damage.

(t)(6) "Moving cables." Energized cables that are to be moved shall be inspected for defects.

(t)(7) "Defective cables." Where a cable in a manhole has one or more abnormalities that could lead to or be an indication of an impending fault, the defective cable shall be de-energized before any employee may work in the manhole, except when service load conditions and a lack of feasible alternatives require that the cable remain energized. In that case, employees may enter the manhole provided they are protected from the possible effects of a failure by shields or other devices that are capable of containing the adverse effects of a fault in the joint.

Note: Abnormalities such as oil or compound leaking from cable or joints, broken cable sheaths or joint sleeves, hot localized surface temperatures of cables or joints, or joints that are swollen beyond normal tolerance are presumed to lead to or be an indication of an impending fault.

(t)(8) "Sheath continuity." When work is performed on buried cable or on cable in manholes, metallic sheath continuity shall be maintained or the cable sheath shall be treated as energized.

(u) "Substations." This paragraph provides additional requirements for substations and for work performed in them.

(u)(1) "Access and working space." Sufficient access and working space shall be provided and maintained about electric equipment to permit ready and safe operation and maintenance of such equipment.

Note: Guidelines for the dimensions of access and workspace about electric equipment in substations are contained in American National Standard-National Electrical Safety Code, ANSI C2-1987. Installations meeting the ANSI provisions comply with paragraph (u)(1) of this section. An installation that does not conform to this ANSI standard will, nonetheless, be considered as complying with paragraph (u)(1) of this section if the employer can demonstrate that the installation provides ready and safe access based on the following evidence:

[1] That the installation conforms to the edition of ANSI C2 that was in effect at the time the installation was made,

[2] That the configuration of the installation enables employees to maintain the minimum approach distances required by paragraph (1)(2) of this section while they are working on exposed, energized parts, and

[3] That the precautions taken when work is performed on the installation provide protection equivalent to the protection that would be provided by access and working space meeting ANSI C2-1987.

(u)(2) "Draw-out-type circuit breakers." When draw-out-type circuit breakers are removed or inserted, the breaker shall be in the open position. The control circuit shall also be rendered inoperative, if the design of the equipment permits.

(u)(3) "Substation fences." Conductive fences around substations shall be grounded. When a substation fence is expanded or a section is removed, fence grounding continuity shall be maintained, and bonding shall be used to prevent electrical discontinuity.

(u)(4) "Guarding of rooms containing electric supply equipment"

(u)(4)(i) Rooms and spaces in which electric supply lines or equipments are installed shall meet the requirements of paragraphs (u)(4)(ii) through (u)(4)(v) of this section under the following conditions:

(u)(4)(i)(A) If exposed live parts operating at 50 to 150 volts to ground are located within 8 feet of the ground or other working surface inside the room or space.

(u)(4)(i)(B) If live parts operating at 151 to 600 volts and located within 8 feet of the ground or other working surface inside the room or space are guarded only by location, as permitted under paragraph (u)(5)(i) of this section, or

(u)(4)(i)(C) If live parts operating at more than 600 volts are located within the room or space, unless:

(u)(4)(i)(C)(1) The live parts are enclosed within grounded, metal-enclosed equipment whose only openings are designed so that foreign objects inserted in these openings will be deflected from energized parts, or

(u)(4)(i)(C)(2) The live parts are installed at a height above ground and any other working surface that provides protection at the voltage to which they are energized corresponding to the protection provided by an 8-foot height at 50 volts.

(u)(4)(ii) The rooms and spaces shall be so enclosed within fences, screens, partitions, or walls as to minimize the possibility that unqualified persons will enter.

(u)(4)(iii) Signs warning unqualified persons to keep out shall be displayed at entrances to the rooms and spaces.

(u)(4)(iv) Entrances to rooms and spaces that are not under the observation of an attendant shall be kept locked.

(u)(4)(v)   Unqualified persons may not enter the rooms or spaces while the electric supply lines or equipment are energized.

(u)(5)   "Guarding of energized parts"

(u)(5)(i)   Guards shall be provided around all live parts operating at more than 150 volts to ground without an insulating covering, unless the location of the live parts gives sufficient horizontal or vertical or a combination of these clearances to minimize the possibility of accidental employee contact.

Note: Guidelines for the dimensions of clearance distances about electric equipment in substations are contained in American National Standard-National Electrical Safety Code, ANSI C2-1987. Installations meeting the ANSI provisions comply with paragraph (u)(5)(i) of this section. An installation that does not conform to this ANSI standard will, nonetheless, be considered as complying with paragraph (u)(5)(i) of this section if the employer can demonstrate that the installation provides sufficient clearance based on the following evidence:

[1] That the installation conforms to the edition of ANSI C2 that was in effect at the time the installation was made,

[2] That each employee is isolated from energized parts at the point of closest approach, and

[3] That the precautions taken when work is performed on the installation provide protection equivalent to the protection that would be provided by horizontal and vertical clearances meeting ANSI C2-1987.

(u)(5)(ii)   Except for fuse replacement and other necessary access by qualified persons, the guarding of energized parts within a compartment shall be maintained during operation and maintenance functions to prevent accidental contact with energized parts and to prevent tools or other equipment from being dropped on energized parts.

(u)(5)(iii)   When guards are removed from energized equipment, barriers shall be installed around the work area to prevent employees who are not working on the equipment, but who are in the area, from contacting the exposed live parts.

(u)(6)   "Substation entry"

(u)(6)(i)   Upon entering an attended substation, each employee other than those regularly working in the station shall report his or her presence to the employee in charge in order to receive information on special system conditions affecting employee safety.

(u)(6)(ii) The job briefing required by paragraph (c) of this section shall cover such additional subjects as the location of energized equipment in or adjacent to the work area and the limits of any de-energized work area.

(v) "Power generation." This paragraph provides additional requirements and related work practices for power generating plants.

(v)(1) "Interlocks and other safety devices"

(v)(1)(i) Interlocks and other safety devices shall be maintained in a safe, operable condition.

(v)(1)(ii) No interlock or other safety device may be modified to defeat its function, except for test, repair, or adjustment of the device.

(v)(2) "Changing brushes." Before exciter or generator brushes are changed while the generator is in service, the exciter or generator field shall be checked to determine whether a ground condition exists. The brushes may not be changed while the generator is energized if a ground condition exists.

(v)(3) "Access and working space." Sufficient access and working space shall be provided and maintained about electric equipment to permit ready and safe operation and maintenance of such equipment.
   Note: Guidelines for the dimensions of access and workspace about electric equipment in generating stations are contained in American National Standard-National Electrical Safety Code, ANSI C2-1987. Installations meeting the ANSI provisions comply with paragraph (v)(3) of this section. An installation that does not conform to this ANSI standard will, nonetheless, be considered as complying with paragraph (v)(3) of this section if the employer can demonstrate that the installation provides ready and safe access based on the following evidence:

[1] That the installation conforms to the edition of ANSI C2 that was in effect at the time the installation was made,

[2] That the configuration of the installation enables employees to maintain the minimum approach distances required by paragraph (1)(2) of this section while they are working on exposed, energized parts, and

[3] That the precautions taken when work is performed on the installation provide protection equivalent to the protection that would be provided by access and working space meeting ANSI C2-1987.

(v)(4) "Guarding of rooms containing electric supply equipment"

(v)(4)(i) Rooms and spaces in which electric supply lines or equipment are installed shall meet the requirements of paragraphs (v)(4)(ii) through (v)(4)(v) of this section under the following conditions:

(v)(4)(i)(A) If exposed live parts operating at 50 to 150 volts to ground are located within 8 feet of the ground or other working surface inside the room or space.

(v)(4)(i)(B) If live parts operating at 151 to 600 volts and located within 8 feet of the ground or other working surface inside the room or space are guarded only by location, as permitted under paragraph (v)(5)(i) of this section, or

(v)(4)(i)(C) If live parts operating at more than 600 volts are located within the room or space, unless:

(v)(4)(i)(C)(1) The live parts are enclosed within grounded, metal-enclosed equipment whose only openings are designed so that foreign objects inserted in these openings will be deflected from energized parts, or

(v)(4)(i)(C)(2) The live parts are installed at a height above ground and any other working surface that provides protection at the voltage to which they are energized corresponding to the protection provided by an 8-foot height at 50 volts.

(v)(4)(ii) The rooms and spaces shall be so enclosed within fences, screens, partitions, or walls as to minimize the possibility that unqualified persons will enter.

(v)(4)(iii) Signs warning unqualified persons to keep out shall be displayed at entrances to the rooms and spaces.

(v)(4)(iv) Entrances to rooms and spaces that are not under the observation of an attendant shall be kept locked.

(v)(4)(v) Unqualified persons may not enter the rooms or spaces while the electric supply lines or equipment are energized.

(v)(5) "Guarding of energized parts"

(v)(5)(i) Guards shall be provided around all live parts operating at more than 150 volts to ground without an insulating covering, unless the location of the live parts gives sufficient horizontal or vertical or a combination of these clearances to minimize the possibility of accidental employee contact.

Note: Guidelines for the dimensions of clearance distances about electric equipment in generating stations are contained in American National Standard-National Electrical Safety Code, ANSI C2-1987. Installations meeting the ANSI provisions comply with paragraph (v)(5)(i) of this section. An installation that does not conform to this ANSI standard will, nonetheless, be considered as complying with paragraph (v)(5)(i) of this section if the employer can demonstrate that the installation provides sufficient clearance based on the following evidence:

[1] That the installation conforms to the edition of ANSI C2 that was in effect at the time the installation was made,

[2] That each employee is isolated from energized parts at the point of closest approach, and

[3] That the precautions taken when work is performed on the installation provide protection equivalent to the protection that would be provided by horizontal and vertical clearances meeting ANSI C2-1987.

(v)(5)(ii) Except for fuse replacement and other necessary access by qualified persons, the guarding of energized parts within a compartment shall be maintained during operation and maintenance functions to prevent accidental contact with energized parts and to prevent tools or other equipment from being dropped on energized parts.

(v)(5)(iii) When guards are removed from energized equipment, barriers shall be installed around the work area to prevent employees who are not working on the equipment, but who are in the area, from contacting the exposed live parts.

(v)(6) "Water or steam spaces." The following requirements apply to work in water and steam spaces associated with boilers:

(v)(6)(i) A designated employee shall inspect conditions before work is permitted and after its completion. Eye protection, or full face protection if necessary, shall be worn at all times when condenser, heater, or boiler tubes are being cleaned.

(v)(6)(ii) Where it is necessary for employees to work near tube ends during cleaning, shielding shall be installed at the tube ends.

(v)(7) "Chemical cleaning of boilers and pressure vessels." The following requirements apply to chemical cleaning of boilers and pressure vessels:

(v)(7)(i) Areas where chemical cleaning is in progress shall be cordoned off to restrict access during cleaning. If flammable liquids, gases, or vapors or combustible materials will be used or might be produced during the cleaning process, the following requirements also apply:

(v)(7)(i)(A) The area shall be posted with signs restricting entry and warning of the hazards of fire and explosion; and

(v)(7)(i)(B) Smoking, welding, and other possible ignition sources are prohibited in these restricted areas.

(v)(7)(ii) The number of personnel in the restricted area shall be limited to those necessary to accomplish the task safely.

(v)(7)(iii) There shall be ready access to water or showers for emergency use.
Note: See § 1910.141 of this Part for requirements that apply to the water supply and to washing facilities.

(v)(7)(iv) Employees in restricted areas shall wear protective equipment meeting the requirements of Subpart I of this Part and including, but not limited to, protective clothing, boots, goggles, and gloves.

(v)(8) "Chlorine systems"

(v)(8)(i) Chlorine system enclosures shall be posted with signs restricting entry and warning of the hazard to health and the hazards of fire and explosion.
Note: See Subpart Z of this Part for requirements necessary to protect the health of employees from the effects of chlorine.

(v)(8)(ii) Only designated employees may enter the restricted area. Additionally, the number of personnel shall be limited to those necessary to accomplish the task safely.

(v)(8)(iii) Emergency repair kits shall be available near the shelter or enclosure to allow for the prompt repair of leaks in chlorine lines, equipment, or containers.

(v)(8)(iv) Before repair procedures are started, chlorine tanks, pipes, and equipment shall be purged with dry air and isolated from other sources of chlorine.

(v)(8)(v) The employer shall ensure that chlorine is not mixed with materials that would react with the chlorine in a dangerously exothermic or other hazardous manner.

(v)(9) "Boilers"

(v)(9)(i) Before internal furnace or ash hopper repair work is started, overhead areas shall be inspected for possible falling objects. If the hazard of falling objects exists, overhead protection such as planking or nets shall be provided.

(v)(9)(ii) When opening an operating boiler door, employees shall stand clear of the opening of the door to avoid the heat blast and gases which may escape from the boiler.

(v)(10) "Turbine generators"

(v)(10)(i) Smoking and other ignition sources are prohibited near hydrogen or hydrogen sealing systems, and signs warning of the danger of explosion and fire shall be posted.

(v)(10)(ii) Excessive hydrogen makeup or abnormal loss of pressure shall be considered as an emergency and shall be corrected immediately.

(v)(10)(iii) A sufficient quantity of inert gas shall be available to purge the hydrogen from the largest generator.

(v)(11) "Coal and ash handling"

(v)(11)(i) Only designated persons may operate railroad equipment.

(v)(11)(ii) Before a locomotive or locomotive crane is moved, a warning shall be given to employees in the area.

(v)(11)(iii) Employees engaged in switching or dumping cars may not use their feet to line up drawheads.

(v)(11)(iv) Drawheads and knuckles may not be shifted while locomotives or cars are in motion.

(v)(11)(v)   When a railroad car is stopped for unloading, the car shall be secured from displacement that could endanger employees.

(v)(11)(vi)   An emergency means of stopping dump operations shall be provided at railcar dumps.

(v)(11)(vii)   The employer shall ensure that employees who work in coal- or ash-handling conveyor areas are trained and knowledgeable in conveyor operation and in the requirements of paragraphs (v)(11)(viii) through (v)(11)(xii) of this section.

(v)(11)(viii)   Employees may not ride a coal- or ash-handling conveyor belt at any time. Employees may not cross over the conveyor belt, except at walkways, unless the conveyor's energy source has been de-energized and has been locked out or tagged in accordance with paragraph (d) of this section.

(v)(11)(ix)   A conveyor that could cause injury when started may not be started until personnel in the area are alerted by a signal or by a designated person that the conveyor is about to start.

(v)(11)(x)   If a conveyor that could cause injury when started is automatically controlled or is controlled from a remote location, an audible device shall be provided that sounds an alarm that will be recognized by each employee as a warning that the conveyor will start and that can be clearly heard at all points along the conveyor where personnel may be present. The warning device shall be actuated by the device starting the conveyor and shall continue for a period of time before the conveyor starts that is long enough to allow employees to move clear of the conveyor system. A visual warning may be used in place of the audible device if the employer can demonstrate that it will provide an equally effective warning in the particular circumstances involved.

Exception: If the employer can demonstrate that the system's function would be seriously hindered by the required time delay, warning signs may be provided in place of the audible-warning device. If the system was installed before January 31, 1995, warning signs may be provided in place of the audible-warning device until such time as the conveyor or its control system is rebuilt or rewired. These warning signs shall be clear, concise, and legible and shall indicate that conveyors and allied equipment may be started at any time, that danger exists, and the personnel must keep clear. These warning signs shall be provided along the conveyor at areas not guarded by position or location.

(v)(11)(xi) Remotely and automatically controlled conveyors, and conveyors that have operating stations which are not manned or which are beyond voice and visual contact from drive areas, loading areas, transfer points, and other locations on the conveyor path not guarded by location, position, or guards shall be furnished with emergency stop buttons, pull cords, limit switches, or similar emergency stop devices. However, if the employer can demonstrate that the design, function, and operation of the conveyor do not expose an employee to hazards, an emergency stop device is not required.

(v)(11)(xi)(A) Emergency stop devices shall be easily identifiable in the immediate vicinity of such locations.

(v)(11)(xi)(B) An emergency stop device shall act directly on the control of the conveyor involved and may not depend on the stopping of any other equipment.

(v)(11)(xi)(C) Emergency stop devices shall be installed so that they cannot be overridden from other locations.

(v)(11)(xii) Where coal-handling operations may produce a combustible atmosphere from fuel sources or from flammable gases or dust, sources of ignition shall be eliminated or safely controlled to prevent ignition of the combustible atmosphere.

Note: Locations that are hazardous because of the presence of combustible dust are classified as Class II hazardous locations. See § 1910.307 of this Part.

(v)(11)(xiii) An employee may not work on or beneath overhanging coal in coal bunkers, coal silos, or coal-storage areas, unless the employee is protected from all hazards posed by shifting coal.

(v)(11)(xiv) An employee entering a bunker or silo to dislodge the contents shall wear a body harness with lifeline attached. The lifeline shall be secured to a fixed support outside the bunker and shall be attended at all times by an employee located outside the bunker or facility.

(v)(12) "Hydroplants and equipment." Employees working on or close to water gates, valves, intakes, forebays, flumes, or other locations where increased or decreased water flow or levels may pose a significant hazard shall be warned and shall vacate such dangerous areas before water-flow changes are made

(w) "Special conditions"

(w)(1) "Capacitors." The following additional requirements apply to work on capacitors and on lines connected to capacitors

Note: See paragraphs (m) and (n) of this section for requirements pertaining to the de-energizing and grounding of capacitor installations.

(w)(1)(i) Before employees work on capacitors, the capacitors shall be disconnected from energized sources and, after a wait of at least 5 minutes from the time of disconnection, short-circuited.

(w)(1)(ii) Before the units are handled, each unit in series-parallel capacitor banks shall be short-circuited between all terminals and the capacitor case or its rack. If the cases of capacitors are on ungrounded substation racks, the racks shall be bonded to ground.

(w)(1)(iii) Any line to which capacitors are connected shall be short-circuited before it is considered de-energized.

(w)(2) "Current transformer secondaries." The secondary of a current transformer may not be opened while the transformer is energized. If the primary of the current transformer cannot be de-energized before work is performed on an instrument, a relay, or other section of a current transformer secondary circuit, the circuit shall be bridged so that the current transformer secondary will not be opened.

(w)(3) "Series streetlighting"

(w)(3)(i) If the open-circuit voltage exceeds 600 volts, the series streetlighting circuit shall be worked in accordance with paragraph (q) or (t) of this section, as appropriate.

(w)(3)(ii) A series loop may only be opened after the streetlighting transformer has been de-energized and isolated from the source of supply or after the loop is bridged to avoid an open-circuit condition.

(w)(4) "Illumination." Sufficient illumination shall be provided to enable the employee to perform the work safely.

(w)(5) "Protection against drowning"

(w)(5)(i) Whenever an employee may be pulled or pushed or may fall into water where the danger of drowning exists, the employee shall be provided with and shall use U.S. Coast Guard approved personal flotation devices.

(w)(5)(ii)   Each personal flotation device shall be maintained in safe condition and shall be inspected frequently enough to ensure that it does not have rot, mildew, water saturation, or any other condition that could render the device unsuitable for use.

(w)(5)(iii)   An employee may cross streams or other bodies of water only if a safe means of passage, such as a bridge, is provided.

(w)(6)   "Employee protection in public work areas"

(w)(6)(i)   Traffic control signs and traffic control devices used for the protection of employees shall meet the requirements of § 1926.200(g)(2) of this Chapter.

(w)(6)(ii)   Before work is begun in the vicinity of vehicular or pedestrian traffic that may endanger employees, warning signs or flags and other traffic control devices shall be placed in conspicuous locations to alert and channel approaching traffic.

(w)(6)(iii)   Where additional employee protection is necessary, barricades shall be used.

(w)(6)(iv)   Excavated areas shall be protected with barricades.

(w)(6)((v)   At night, warning lights shall be prominently displayed.

(w)(7)   "Backfeed." If there is a possibility of voltage backfeed from sources of cogeneration or from the secondary system (for example, backfeed from more than one energized phase feeding a common load), the requirements of paragraph (l) of this section apply if the lines or equipment are to be worked as energized, and the requirements of paragraphs (m) and (n) of this section apply if the lines or equipment are to be worked as de-energized.

(w)(8)   "Lasers." Laser equipment shall be installed, adjusted, and operated in accordance with § 1926.54 of this Chapter.

(w)(9)   "Hydraulic fluids." Hydraulic fluids used for the insulated sections of equipment shall provide insulation for the voltage involved.

(x) Definitions

- *Affected employee.* An employee whose job requires him or her to operate or use a machine or equipment on which servicing or maintenance is being performed under lockout or tagout, or whose job requires him or her to work in an area in which such servicing or maintenance is being performed.

- *Attendant.* An employee assigned to remain immediately outside the entrance to an enclosed or other space to render assistance as needed to employees inside the space.

- *Authorized employee.* An employee who locks out or tags out machines or equipment in order to perform servicing or maintenance on that machine or equipment. An affected employee becomes an authorized employee when that employee's duties include performing servicing or maintenance covered under this section.

- *Automatic circuit recloser.* A self-controlled device for interrupting and reclosing an alternating current circuit with a predetermined sequence of opening and reclosing followed by resetting, hold-closed, or lockout operation.

- *Barricade.* A physical obstruction such as tapes, cones, or A-frame type wood or metal structures intended to provide a warning about and to limit access to a hazardous area.

- *Barrier.* A physical obstruction which is intended to prevent contact with energized lines or equipment or to prevent unauthorized access to a work area.

- *Bond.* The electrical interconnection of conductive parts designed to maintain a common electrical potential.

- *Bus.* A conductor or a group of conductors that serve as a common connection for two or more circuits.

- *Bushing.* An insulating structure, including a through conductor or providing a passageway for such a conductor, with provision for mounting on a barrier, conducting or otherwise, for the purposes of insulating the conductor from the barrier and conducting current from one side of the barrier to the other.

- *Cable.* A conductor with insulation, or a stranded conductor with or without insulation and other coverings (single-conductor cable), or a combination of conductors insulated from one another (multiple-conductor cable).

- *Cable sheath.* A conductive protective covering applied to cables.
   Note: A cable sheath may consist of multiple layers of which one or more is conductive.

- *Circuit.* A conductor or system of conductors through which an electric current is intended to flow.

- *Clearance (between objects).* The clear distance between two objects measured surface to surface.

- *Clearance (for work).* Authorization to perform specified work or permission to enter a restricted area.

- *Communication lines.* (*See* Lines, communication.)

- *Conductor.* A material, usually in the form of a wire, cable, or bus bar, used for carrying an electric current.

- *Covered conductor.* A conductor covered with a dielectric having no rated insulating strength or having a rated insulating strength less than the voltage of the circuit in which the conductor is used.

- *Current-carrying part.* A conducting part intended to be connected in an electric circuit to a source of voltage. Noncurrent-carrying parts are those not intended to be so connected.

- *De-energized.* Free from any electrical connection to a source of potential difference and from electric charge; not having a potential different from that of the earth.

  Note: The term is used only with reference to current-carrying parts, which are sometimes energized (alive).

- *Designated employee (designated person).* An employee (or person) who is designated by the employer to perform specific duties under the terms of this section and who is knowledgeable in the construction and operation of the equipment and the hazards involved.

- *Electric line truck.* A truck used to transport personnel, tools, and material for electric supply line work.

- *Electric supply equipment.* Equipment that produces, modifies, regulates, controls, or safeguards a supply of electric energy.

- *Electric supply lines.* (*See* Lines, electric supply.)

- *Electric utility.* An organization responsible for the installation, operation, or maintenance of an electric supply system.

- *Enclosed space.* A working space, such as a manhole, vault, tunnel, or shaft, that has a limited means of egress or entry, that is designed for periodic employee entry under normal operating conditions, and that under normal conditions does not contain a hazardous atmosphere, but that may contain a hazardous atmosphere under abnormal conditions.

  Note: Spaces that are enclosed but not designed for employee entry under normal operating conditions are not considered to be enclosed spaces for the purposes of this section. Similarly, spaces that are enclosed and that are expected to contain a hazardous atmosphere are not considered to be enclosed spaces for the purposes of this section. Such spaces meet the definition of permit spaces in § 1910.146 of this Part, and entry into them must be performed in accordance with that standard.

- *Energized* (*alive, live*). Electrically connected to a source of potential difference, or electrically charged so as to have a potential significantly different from that of earth in the vicinity.

- *Energy isolating device.* A physical device that prevents the transmission or release of energy, including, but not limited to, the following: a manually operated electric circuit breaker, a disconnect switch, a manually operated switch, a slide gate, a slip blind, a line valve, blocks, and any similar device with a visible indication of the position of the device. (Push buttons, selector switches, and other control-circuit-type devices are not energy isolating devices.)

- *Energy source.* Any electrical, mechanical, hydraulic, pneumatic, chemical, nuclear, thermal, or other energy source that could cause injury to personnel.

- *Equipment* (*electric*). A general term including material, fittings, devices, appliances, fixtures, apparatus, and the like used as part of or in connection with an electrical installation.

- *Exposed.* Not isolated or guarded.

- *Ground.* A conducting connection, whether intentional or accidental, between an electric circuit or equipment and the earth, or to some conducting body that serves in place of the earth.

- *Grounded.* Connected to earth or to some conducting body that serves in place of the earth.

- *Guarded.* Covered, fenced, enclosed, or otherwise protected, by means of suitable covers or casings, barrier rails or screens, mats, or platforms, designed to minimize the possibility, under normal conditions, of dangerous approach or accidental contact by persons or objects.

  Note: Wires which are insulated, but not otherwise protected, are not considered as guarded.

- *Hazardous atmosphere.* Means an atmosphere that may expose employees to the risk of death, incapacitation, impairment of ability to self-rescue (that is, escape unaided from an enclosed space), injury, or acute illness from one or more of the following causes:

  (x)(1) Flammable gas, vapor, or mist in excess of 10 percent of its lower flammable limit (LFL);

  (x)(2) Airborne combustible dust at a concentration that meets or exceeds its LFL;

  Note: This concentration may be approximated as a condition in which the dust obscures vision at a distance of 5 feet (1.52 m) or less.

(x)(3) Atmospheric oxygen concentration below 19.5 percent or above 23.5 percent;

(x)(4) Atmospheric concentration of any substance for which a dose or a permissible exposure limit is published in Subpart G, "Occupational Health and Environmental Control," or in Subpart Z, "Toxic and Hazardous Substances," of this Part and which could result in employee exposure in excess of its dose or permissible exposure limit;

Note: An atmospheric concentration of any substance that is not capable of causing death, incapacitation, impairment of ability to self-rescue, injury, or acute illness due to its health effects is not covered by this provision.

(x)(5) Any other atmospheric condition that is immediately dangerous to life or health.

Note: For air contaminants for which OSHA has not determined a dose or permissible exposure limit, other sources of information, such as Material Safety Data Sheets that comply with the Hazard Communication Standard, § 1910.1200 of this Part, published information, and internal documents can provide guidance in establishing acceptable atmospheric conditions.

- *High-power tests.* Tests in which fault currents, load currents, magnetizing currents, and line-dropping currents are used to test equipment, either at the equipment's rated voltage or at lower voltages.

- *High-voltage tests.* Tests in which voltages of approximately 1000 volts are used as a practical minimum and in which the voltage source has sufficient energy to cause injury.

- *High wind.* A wind of such velocity that the following hazards would be present:

  [1] An employee would be exposed to being blown from elevated locations, or

  [2] An employee or material-handling equipment could lose control of material being handled, or

  [3] An employee would be exposed to other hazards not controlled by the standard involved.

Note: Winds exceeding 40 miles per hour (64.4 kilometers per hour), or 30 miles per hour (48.3 kilometers per hour) if material handling is involved, are normally considered as meeting this criteria unless precautions are taken to protect employees from the hazardous effects of the wind.

- *Immediately dangerous to life or health (IDLH)*. Means any condition that poses an immediate or delayed threat to life or that would cause irreversible adverse health effects or that would interfere with an individual's ability to escape unaided from a permit space.

  Note: Some materials—hydrogen fluoride gas and cadmium vapor, for example—may produce immediate transient effects that, even if severe, may pass without medical attention, but are followed by sudden, possibly fatal collapse 12–72 hours after exposure. The victim "feels normal" from recovery from transient effects until collapse. Such materials in hazardous quantities are considered to be "immediately" dangerous to life or health.

- *Insulated.* Separated from other conducting surfaces by a dielectric (including air space) offering a high resistance to the passage of current.

  Note: When any object is said to be insulated, it is understood to be insulated for the conditions to which it is normally subjected. Otherwise, it is, within the purpose of this section, uninsulated.

- *Insulation (cable).* That which is relied upon to insulate the conductor from other conductors or conducting parts or from ground.

- *Line-clearance tree trimmer.* An employee who, through related training or on-the-job experience or both, is familiar with the special techniques and hazards involved in line-clearance tree trimming.

  Note 1: An employee who is regularly assigned to a line-clearance tree-trimming crew and who is undergoing on-the-job training and who, in the course of such training, has demonstrated an ability to perform duties safely at his or her level of training and who is under the direct supervision of a line-clearance tree trimmer is considered to be a line-clearance tree trimmer for the performance of those duties.

  Note 2: A line-clearance tree trimmer is not considered to be a "qualified employee" under this section unless he or she has the training required for a qualified employee under paragraph (a)(2)(ii) of this section. However, under the electrical safety-related work practices standard in Subpart S of this Part, a line-clearance tree trimmer is considered to be a "qualified employee." Tree trimming performed by such "qualified employees" is not subject to the electrical safety-related work practice requirements contained in § 1910.331 through § 1910.335 of this Part. [*See also* the note following § 1910.332(b)(3) of this Part for information regarding the training an employee must have to be considered a qualified employee under § 1910.331 through § 1910.335 of this part.]

- *Line-clearance tree trimming.* The pruning, trimming, repairing, maintaining, removing, or clearing of trees or the cutting of brush that is within 10 feet (305 cm) of electric supply lines and equipment.

- *Lines.* [1] *Communication lines.* The conductors and their supporting or containing structures which are used for public or private signal or communication service, and which operate at potentials not exceeding 400 volts to ground or 750 volts between any two points of the circuit, and the transmitted power of which does not exceed 150 watts. If the lines are operating at less than 150 volts, no limit is placed on the transmitted power of the system. Under certain conditions, communication cables may include communication circuits exceeding these limitations where such circuits are also used to supply power solely to communication equipment.

  Note: Telephone, telegraph, railroad signal, data, clock, fire, police alarm, cable television, and other systems conforming to this definition are included. Lines used for signaling purposes, but not included under this definition, are considered as electric supply lines of the same voltage.

  [2] *Electric supply lines.* Conductors used to transmit electric energy and their necessary supporting or containing structures. Signal lines of more than 400 volts are always supply lines within this section, and those of less than 400 volts are considered as supply lines, if so run and operated throughout.

- *Manhole.* A subsurface enclosure, which personnel may enter and, which is used for the purpose of installing, operating, and maintaining submersible equipment or cable.

- *Manhole steps.* A series of steps individually attached to or set into the walls of a manhole structure.

- *Minimum approach distance.* The closest distance an employee is permitted to approach an energized or a grounded object.

- *Qualified employee (qualified person).* One knowledgeable in the construction and operation of the electric power generation, transmission, and distribution equipment involved, along with the associated hazards.

  Note 1: An employee must have the training required by paragraph (a)(2)(ii) of this section in order to be considered a qualified employee.

  Note 2: Except under paragraph (g)(2)(v) of this section, an employee who is undergoing on-the-job training and who, in the course of such training, has demonstrated an ability to perform duties safely at his or her level of training and who is under the direct supervision of a qualified person is considered to be a qualified person for the performance of those duties.

- *Step bolt.* A bolt or rung attached at intervals along a structural member and used for foot placement during climbing or standing.

- *Switch.* A device for opening and closing or for changing the connection of a circuit. In this section, a switch is understood to be manually operable, unless otherwise stated.

- *System operator.* A qualified person designated to operate the system or its parts.

- *Vault.* An enclosure, above or below ground, which personnel may enter and which is used for the purpose of installing, operating, or maintaining equipment or cable.

- *Vented vault.* A vault that has provision for air changes using exhaust flue stacks and low-level air intakes operating on differentials of pressure and temperature providing for airflow which precludes a hazardous atmosphere from developing.

- *Voltage.* The effective (rms) potential difference between any two conductors or between a conductor and ground. Voltages are expressed in nominal values unless otherwise indicated. The nominal voltage of a system or circuit is the value assigned to a system or circuit of a given voltage class for the purpose of convenient designation. The operating voltage of the system may vary above or below this value.

[59 FR 40672, Aug. 9, 1994; 59 FR 51672, Oct. 12, 1994]

Appendix A to § 1910.269—Flowcharts

This appendix presents information, in the form of flowcharts, that illustrates the scope and application of § 1910.269. This appendix addresses the interface between § 1910.269 and Subpart S of this Part (Electrical), between § 1910.269 and § 1910.146 of this Part (Permit-required confined spaces), and between § 1910.269 and § 1910.147 of this Part [The control of hazardous energy (lockout//tagout)]. These flowcharts provide guidance for employers trying to implement the requirements of § 1910.269 in combination with other General Industry Standards contained in Part § 1910.

Appendix A.1 to §1910.269—Application of §1910.269 and Subpart S of this Part to Electrical Installations

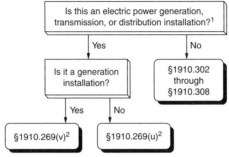

Is this an electric power generation, transmission, or distribution installation?[1]

- Yes → Is it a generation installation?
  - Yes → §1910.269(v)[2]
  - No → §1910.269(u)[2]
- No → §1910.302 through §1910.308

[1]Electrical installation design requirements only. See Appendix 1B for electrical safety-related work practices. Supplementary electric generating equipment that is used to supply a workplace for emergency, standby, or similar purposes only is not considered to be an electric power generation installation.

[2]See Table 1 of Appendix A.2 for requirements that can be met through compliance with Subpart S.

Appendix A.2 to §1910.269—Application of §1910.269 and Subpart S to Electrical Safety-Related Work Practices

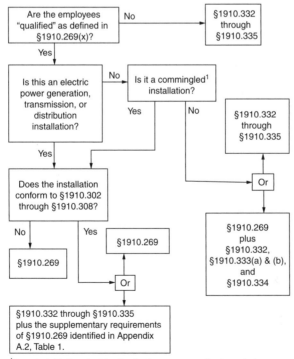

Are the employees "qualified" as defined in §1910.269(x)?

- No → §1910.332 through §1910.335
- Yes → Is this an electric power generation, transmission, or distribution installation?
  - No → Is it a commingled[1] installation?
    - Yes →
    - No → §1910.332 through §1910.335
  - Yes → Does the installation conform to §1910.302 through §1910.308?
    - No → §1910.269
    - Yes → §1910.269

Or → §1910.332 through §1910.335 plus the supplementary requirements of §1910.269 identified in Appendix A.2, Table 1.

Or → §1910.269 plus §1910.332, §1910.333(a) & (b), and §1910.334

[1]Commingled to the extent that the electric power generation, transmission, or distribution installation poses the greater hazard.

Compliance with Subpart S is considered as compliance with § 1910.269[1]	Paragraphs that apply regardless of compliance with Subpart S
(d) electric shock hazards only................	$(a)(2)^2$ and $(a)(3)^2$.
(h)(3)..	$(b)^2$.
(i)(2)...	$(c)^2$.
(k)...	(d), other than electric shock hazards.
(l)(1) through (l)(4), (l)(6)(i), and	
(l)(8) through (l)(10)...............................	(e).
(m)...	(f).
(p)(4)..	(g).
(s)(2)..	(h)(1) and (h)(2).
(u)(1) and (u)(3) through (u)(5)...............	$(i)(3)^2$ and $(i)(4)^2$.
(v)(3) through (v)(5)...............................	(j)2.
(w)(1) and (w)(7)...................................	$(l)(5)^2$, $(l)(6)(iii)^2$, $(l)(6)(iii)^2$, and $(l)(7)^2$.
	$(n)^2$.
	$(o)^2$.
	(p)(1) through (p)(3).
	$(q)^2$
	(r).
	(s)(1).
	$(t)^2$.
	$(u)(2)^2$ and $(u)(6)^2$.
	(v)(1), $(v)(2)^2$, and (v)(6) through (v)(12).
	(w)(2) through $(w)(6)^2$, (w)(8), and $(w)(9)^2$.

[1]If the electrical installation meets the requirements of §§ 1910.332 through 1910.308 of this Part, then the electrical installation and any associated electrical safety-related work practices conforming to §§ 1910.332 through 1910.335 of this Part are considered to comply with these provisions of § 1910.269 of the Part.

[2]These provisions include electrical safety requirements that must be met regardless of compliance with Subpart S of this Part.

Appendix A.3 to §1910.269—Application of §1910.269 and Subpart S of this Part to Tree-Trimming Operations

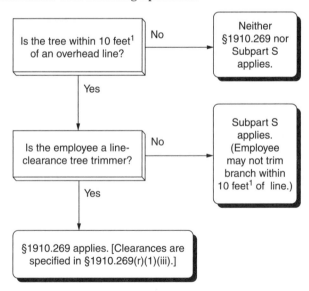

[1]10 feet plus 4 inches for every 10 kilovolts over 50 kilovolts.

Appendix A.4 to §1910.269—Application of §§ 1910.147, § 1910.269 and § 1910.333 to Hazardous Energy Control Procedures (Lockout/Tagout)

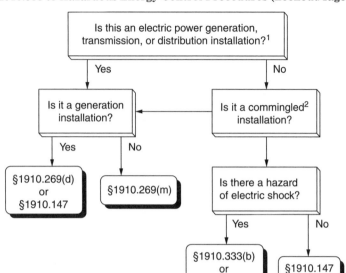

[1] If the installation conforms to §1910.303 through §1910.308, the lockout and tagging procedures of 1910.333(b) may be followed for electric shock hazards.

[2] Commingled to the extent that the electric power generation, transmission, or distribution installation poses the greater hazard.

[3] §1910.333(b)(2)(iii)(D) and (b)(2)(iv)(B) still apply.

Appendix B to § 1910.269—Working on Exposed Energized Parts

I. Introduction

Electric transmission and distribution line installations have been designed to meet National Electrical Safety Code (NESC), ANSI C2, requirements and to provide the level of line outage performance required by system reliability criteria. Transmission and distribution lines are also designed to withstand the maximum overvoltages expected to be impressed on the system. Such overvoltages can be caused by such conditions as switching surges, faults, or lightning. Insulator design and lengths and the clearances to structural parts (which, for low voltage through extra-high voltage, or EHV, facilities, are generally based on the performance of the line as a result of contamination of the insulation or during storms) have, over the years, come closer to the minimum approach distances used by workers (which are generally based on nonstorm conditions). Thus, as minimum approach (working) distances

Appendix A.5 to § 1910.269—Application of §§ 1910.146 and § 1910.269 to Permit-Required Confined Spaces

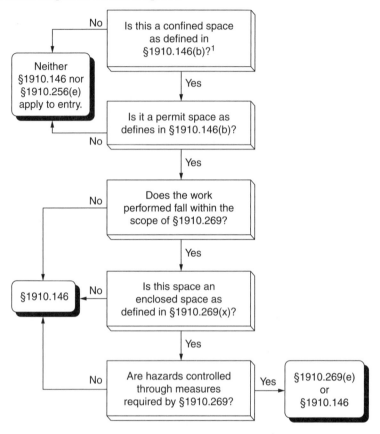

[1]See §1910.146(c) for general non-entry requirements that apply to all confined spaces.

and structural distances (clearances) converge, it is increasingly important that basic considerations for establishing safe approach distances for performing work be understood by the designers and the operating and maintenance personnel involved.

The information in this appendix will assist employers in complying with the minimum approach distance requirements contained in paragraphs (l)(2) and (q)(3) of this section. The technical criteria and methodology presented herein is mandatory for employers using reduced minimum approach distances as permitted in Tables R.7 and R.8. This appendix is intended to provide essential background information and technical criteria for the development or modification, if possible, of the safe minimum approach distances for electric transmission and distribution live-line work. The development of these safe distances must be undertaken by persons knowledgeable in the techniques discussed in this appendix and competent in the field of electric transmission and distribution system design.

II. General

A. Definitions. The following definitions from § 1910.269(x) relate to work on or near transmission and distribution lines and equipment and the electrical hazards they present:

Exposed. Not isolated or guarded.

Guarded. Covered, fenced, enclosed, or otherwise protected, by means of suitable covers or casings, barrier rails or screens, mats, of platforms, designed to minimize the possibility, under normal conditions, of dangerous approach or accidental contact by persons or objects.

Note: Wires which are insulated, but not otherwise protected, are not considered as guarded.

Insulated. Separated from other conducting surfaces by a dielectric (including air space) offering a high resistance to the passage of current.

Note: When any object is said to be insulated, it is understood to be insulated for the conditions to which it is normally subjected. Otherwise, it is, within the purpose of this section, uninsulated.

B. Installations energized at 50 to 300 volts. The hazards posed by installations energized at 50 to 300 volts are the same as those found in many other workplaces. That is not to say that there is no hazard, but the complexity of electrical protection required does not compare to that required for high voltage systems. The employee must avoid contact with the exposed parts, and the protective equipment used (such as rubber insulating gloves) must provide insulation for the voltages involved.

C. Exposed energized parts over 300 volts AC. Tables R.6, R.7, and R.8 of § 1910.269 provide safe approach and working distances in the vicinity of energized electric apparatus so that work can be done safely without risk of electrical flashover.

The working distances must withstand the maximum transient overvoltage that can reach the work site under the working conditions and practices in use. Normal system design may provide or include a means to control transient overvoltages, or temporary devices may be employed to achieve the same result. The use of technically correct practices or procedures to control overvoltages (for example, portable gaps or preventing the automatic control from initiating breaker reclosing) enables line design and operation to be based on reduced transient overvoltage values. Technical information for U.S. electrical systems indicates that current design provides for the following maximum transient overvoltage values (usually produced by switching surges): 362 kV and less—3.0 per unit; 552 kV—2.4 per unit; 800 kV—2.0 per unit.

Additional discussion of maximum transient overvoltages can be found in paragraph IV.A.2, later in this appendix.

III. Determination of the electrical component of minimum approach distances

A. Voltages of 1.1 kV to 72.5 kV. For voltages of 1.1 kV to 72.5 kV, the electrical component of minimum approach distances is based on American National Standards Institute (ANSI)/American Institute of Electrical Engineers (AIEE) Standard No. 4, March 1943, Tables III and IV. (AIEE is the predecessor technical society to the Institute of Electrical and Electronic Engineers (IEEE).) These distances are calculated by the following formula:

Equation (1)—For Voltages of 1.1 kV to 72.5 kV

$$D = \frac{(V_{max} \times pu)^{1.63}}{124}$$

where D = electrical component of the minimum approach distance in air in feet
V_{max} = maximum rated line-to-ground rms voltage in kV
pu = maximum transient overvoltage factor in per unit

Source: AIEE Standard No. 4, 1943.
This formula has been used to generate Table 1.

TABLE 1   AC Energized Line-Work Phase-to-Ground Electrical Component of the Minimum Approach Distance—1.1 to 72.5 kV

Maximum anticipated per-unit transient overvoltage	Phase to phase voltage			
	15,000	36,000	46,000	72,500
3.0....................	0.08	0.33	0.49	1.03

NOTE: The distances given (in feet) are for air as the insulating medium and provide no additional clearance for inadvertent movement.

B. Voltages of 72.6 kV to 800 kV.   For voltages of 72.6 to 800 kV, the electrical component of minimum approach distances in based on ANSI/IEEE Standard 516-1987, "IEEE Guide for Maintenance Methods on Energized Power Lines." This standard gives the electrical component of the minimum approach distance based on power frequency rod-gap data, supplemented with transient overvoltage information and a saturation factor for high voltages. The distances listed in ANSI/IEEE Standard 516 have been calculated according to the following formula:

Equation (2)—For Voltages of 72.6 kV to 800 kV

$$D = (C + a)\text{pu } V_{max}$$

where D = electrical component of the minimum approach distance in air in feet
    C = 0.01 to take care of correction factors associated with the variation of gap sparkover with voltage
    a = factor relating to the saturation of air at voltages of 345 kV or higher
    pu = maximum anticipated transient overvoltage, in per unit (p.u.)
    V_{max} = maximum rms system line-to-ground voltage in kilovolts—it should be the "actual" maximum, or the normal highest voltage for the range (for example, 10 percent above the nominal voltage)

Source: Formula developed from ANSI/IEEE Standard No. 516, 1987.

This formula is used to calculate the electrical component of the minimum approach distances in air and is used in the development of Tables 2 and 3.

C. Provisions for inadvertent movement.   The minimum approach distances (working distances) must include an "adder" to compensate for the inadvertent movement of the worker relative to an energized part or the movement of the part relative to the worker. A certain allowance must be made to account for this possible inadvertent movement and to provide the worker with a comfortable and safe zone in which to work. A distance for inadvertent movement (called the "ergonomic component of the minimum approach distance") must be added to the electrical component to determine the total safe minimum approach distances used in live-line work.

One approach that can be used to estimate the ergonomic component of the minimum approach distance is response time-distance analysis. When this technique is used, the total response time to a hazardous

TABLE 2   AC Energized Line-Work Phase-to-Ground Electrical Component of the
Minimum Approach Distance—121 to 242 kV

Maximum anticipated per-unit transient overvoltage	Phase to phase voltage			
	121,000	145,000	169,000	242,000
2.0..................	1.40	1.70	2.00	2.80
2.1..................	1.47	1.79	2.10	2.94
2.2..................	1.54	1.87	2.20	3.08
2.3..................	1.61	1.96	2.30	3.22
2.4..................	1.68	2.04	2.40	3.35
2.5..................	1.75	2.13	2.50	3.50
2.6..................	1.82	2.21	2.60	3.64
2.7..................	1.89	2.30	2.70	3.76
2.8..................	1.96	2.38	2.80	3.92
2.9..................	2.03	2.47	2.90	4.05
3.0..................	2.10	2.55	3.00	4.29

NOTE: The distances given (in feet) are for air as the insulating medium and provide no additional clearance for inadvertent movement.

TABLE 3   AC Energized Line-Work Phase-to-Ground Electrical Component of the
Minimum Approach Distance—362 to 800 kV

Maximum anticipated per-unit transient overvoltage	Phase to phase voltage		
	362,000	552,000	800,000
1.5.....................	. . . .	4.97	8.66
1.6.....................	. . . .	5.46	9.60
1.7.....................	. . . .	5.98	10.60
1.8.....................	. . . .	6.51	11.64
1.9.....................	. . . . . .	7.08	12.73
2.0.....................	4.20	7.68	13.86
2.1.....................	4.41	8.27	. . . .
2.2.....................	4.70	8.87	. . . .
2.3.....................	5.01	9.49	. . . .
2.4.....................	5.34	10.21	. . . .
2.5.....................	5.67	. . . .	. . . .
2.6.....................	6.01	. . . .	. . . .
2.7.....................	6.36	. . . .	. . . .
2.8.....................	6.73	. . . .	. . . .
2.9.....................	7.10	. . . .	. . . .
3.0.....................	7.48	. . . .	. . . .

NOTE: The distances given (in feet) are for air as the insulating medium and provide no additional clearance for inadvertent movement.

incident is estimated and converted to distance travelled. For example, the driver of a car takes a given amount of time to respond to a "stimulus" and stop the vehicle. The elapsed time involved results in a distance being travelled before the car comes to a complete stop. This distance is dependent on the speed of the car at the time the stimulus appears.

In the case of live-line work, the employee must first perceive that he or she is approaching the danger zone. Then, the worker responds to the danger and must decelerate and stop all motion toward the energized part. During the time it takes to stop, a distance will have been traversed. It is this distance that must be added to the electrical component of the minimum approach distance to obtain the total safe minimum approach distance.

At voltages below 72.5 kV, the electrical component of the minimum approach distance is smaller than the ergonomic component. At 72.5 kV the electrical component is only a little more than 1 foot. An ergonomic component of the minimum approach distance is needed that will provide for all the worker's expected movements. The usual live-line work method for these voltages is the use of rubber insulating equipment, frequently rubber gloves. The energized object needs to be far enough away to provide the worker's face with a safe approach distance, as his or her hands and arms are insulated. In this case, 2 feet has been accepted as a sufficient and practical value.

For voltages between 72.6 and 800 kV, there is a change in the work practices employed during energized line work. Generally, live-line tools (hot sticks) are employed to perform work while equipment is energized. These tools, by design, keep the energized part at a constant distance from the employee and thus maintain the appropriate minimum approach distance automatically.

The length of the ergonomic component of the minimum approach distance is also influenced by the location of the worker and by the nature of the work. In these higher voltage ranges, the employees use work methods that more tightly control their movements than when the workers perform rubber glove work. The worker is farther from energized line or equipment and needs to be more precise in his or her movements just to perform the work.

For these reasons, a smaller ergonomic component of the minimum approach distance is needed, and a distance of 1 foot has been selected for voltages between 72.6 and 800 kV.

Table 4 summarizes the ergonomic component of the minimum approach distance for the two voltage ranges.

TABLE 4   Ergonomic Component of Minimum Approach Distance

Voltage range (kV)	Distance (feet)
1.1 to 72.5.......	2.0
72.6 to 800........	1.0

NOTE: This distance must be added to the electrical component of the minimum approach distance to obtain the full minimum approach distance.

D. Bare-hand live-line minimum approach distances. Calculating the strength of phase-to-phase transient overvoltages is complicated by the varying time displacement between overvoltages on parallel conductors (electrodes) and by the varying ratio between the positive and negative voltages on the two electrodes. The time displacement causes the maximum voltage between phases to be less than the sum of the phase-to-ground voltages. The International Electrotechnical Commission (IEC) Technical Committee 28, Working Group 2, has developed the following formula for determining the phase-to-phase maximum transient overvoltage, based on the per unit (pu) of the system nominal voltage phase-to-ground crest:

$$pu(p) = pu(g) + 1.6$$

where pu(g) = pu phase-to-ground maximum transient overvoltage
       pu(p) = pu phase-to-phase maximum transient overvoltage

This value of maximum anticipated transient overvoltage must be used in Equation (2) to calculate the phase-to-phase minimum approach distances for live-line bare-hand work.

E. Compiling the minimum approach distance tables. For each voltage involved, the distance in Table 4 in this appendix has been added to the distance in Tables 1, 2, or 3 in this appendix to determine the resulting minimum approach distances in Tables R.6, R.7, and R.8 in § 1910.269.

F. Miscellaneous correction factors. The strength of an air gap is influenced by the changes in the air medium that forms the insulation. A brief discussion of each factor follows, with a summary at the end.

1. *Dielectric strength of air.* The dielectric strength of air in a uniform electric field at standard atmospheric conditions is approximately 31 kV (crest) per cm at 60 Hz. The disruptive gradient is affected by the air pressure, temperature, and humidity, by the shape, dimensions, and separation of the electrodes, and by the characteristics of the applied voltage (wave shape).

2. *Atmospheric effect.* Flashover for a given air gap is inhibited by an increase in the density (humidity) of the air. The empirically determined electrical strength of a given gap is normally applicable at standard atmospheric conditions (20 deg. C, 101.3 kPa, 11 g/cm^3 humidity).

   The combination of temperature and air pressure that gives the lowest gap flash-over voltage is high temperature and low pressure. These are conditions not likely to occur simultaneously. Low air pressure is generally associated with high humidity, and this causes increased electrical strength. An average air pressure is more likely to be associated with low humidity. Hot and dry working conditions are thus normally associated with reduced electrical strength.

The electrical component of the minimum approach distances in Tables 1, 2, and 3 has been calculated using the maximum transient overvoltages to determine withstand voltages at standard atmospheric conditions.

3. *Altitude.* The electrical strength of an air gap is reduced at high altitude, due principally to the reduced air pressure. An increase of 3 percent per 300 meters in the minimum approach distance for altitudes above 900 meters is required. Table R.10 of § 1910.269 presents the information in tabular form.

Summary.   After taking all these correction factors into account and after considering their interrelationships relative to the air gap insulation strength and the conditions under which live work is performed, one finds that only a correction for altitude need be made. An elevation of 900 meters is established as the base elevation, and the values of the electrical component of the minimum approach distances have been derived with this correction factor in mind. Thus, the values used for elevations below 900 meters are conservative without any change; corrections have to be made only above this base elevation.

IV. Determination of reduced minimum approach distances

A. Factors affecting voltage stress at the work site

1. *System voltage (nominal).* The nominal system voltage range sets the absolute lower limit for the minimum approach distance. The highest value within the range, as given in the relevant table, is selected and used as a reference for per unit calculations.

2. *Transient overvoltages.* Transient overvoltages may be generated on an electrical system by the operation of switches or breakers, by the occurrence of a fault on the line or circuit being worked or on an adjacent circuit, and by similar activities. Most of the overvoltages are caused by switching, and the term "switching surge" is often used to refer generically to all types of overvoltages. However, each overvoltage has an associated transient voltage wave shape. The wave shape arriving at the site and its magnitude vary considerably.

The information used in the development of the minimum approach distances takes into consideration the most common wave shapes; thus, the required minimum approach distances are appropriate for any transient overvoltage level usually found on electric power

TABLE 5   Magnitude of Typical Transient Overvoltages

Cause	Magnitude (per unit)
Energized 200 mile line without closing resistors	3.5
Energized 200 mile line with one step closing resistor..............	2.1
Energized 200 mile line with multi-step resistor........................	2.5
Reclosed with trapped charge one-step resistor.........................	2.2
Opening surge with single restrike...	3.0
Fault initiation unfaulted phase..	2.1
Fault initiation adjacent circuit..	2.5
Fault clearing..	1.7–1.9

SOURCE: ANSI/IEEE Standard No. 516, 1987.

generation, transmission, and distribution systems. The values of the per unit (pu) voltage relative to the nominal maximum voltage are used in the calculation of these distances.

3. *Typical magnitude of overvoltages.* The magnitude of typical transient overvoltages is given in Table 5.

4. *Standard deviation—air-gap withstand.* For each air-gap length, and under the same atmospheric conditions, there is a statistical variation in the breakdown voltage. The probability of the breakdown voltage is assumed to have a normal (Gaussian) distribution. The standard deviation of this distribution varies with the wave shape, gap geometry, and atmospheric conditions. The withstand voltage of the air gap used in calculating the electrical component of the minimum approach distance has been set at three standard deviations (3 σ-sigma[1]) below the critical flashover voltage. (The critical flashover voltage is the crest value of the impulse wave that, under specified conditions, causes flashover on 50 percent of the applications. An impulse wave of three standard deviations below this value, that is, the withstand voltage, has a probability of flashover of approximately 1 in 1000.)

5. *Broken insulators.* Tests have shown that the insulation strength of an insulator string with broken skirts is reduced. Broken units may have lost up to 70 percent of their withstand capacity. Because the insulating capability of a broken unit cannot be determined without testing it, damaged units in an insulator are usually considered to have no insulating value. Additionally, the overall insulating strength of a string with broken units may be further reduced in the presence of a live-line tool alongside it. The number of good units that must be present in a string is based on the maximum overvoltage possible at the work site.

---

[1] σ -Sigma is the symbol for standard deviation.

B. Minimum approach distances based on known maximum anticipated per-unit transient overvoltages

1. *Reduction of the minimum approach distance for AC systems.* When the transient overvoltage values are known and supplied by the employer, Tables R.7 and R.8 of § 1910.269 allow the minimum approach distances from energized parts to be reduced. In order to determine what this maximum overvoltage is, the employer must undertake an engineering analysis of the system. As a result of this engineering study, the employer must provide new live work procedures, reflecting the new minimum approach distances, the conditions and limitations of application of the new minimum approach distances, and the specific practices to be used when these procedures are implemented.

2. *Calculation of reduced approach distance values.* The following method of calculating reduced minimum approach distances is based on ANSI/IEEE Standard 516:

Step 1. Determine the maximum voltage (with respect to a given nominal voltage range) for the energized part.

Step 2. Determine the maximum transient overvoltage (normally a switching surge) that can be present at the work site during work operation.

Step 3. Determine the technique to be used to control the maximum transient overvoltage. (See paragraphs IV.C and IV.D of this appendix.) Determine the maximum voltage that can exist at the work site with that form of control in place and with a confidence level of 3 σ. This voltage is considered to be the withstand voltage for the purpose of calculating the appropriate minimum approach distance.

Step 4. Specify in detail the control technique to be used, and direct its implementation during the course of the work.

Step 5. Using the new value of transient overvoltage in per unit (pu), determine the required phase-to-ground minimum approach distance from Table R.7 or Table R.8 of § 1910.269.

C. Methods of controlling possible transient overvoltage stress found on a system

1. *Introduction.* There are several means of controlling overvoltages that occur on transmission systems. First, the operation of circuit breakers or other switching devices may be modified to reduce switching transient overvoltages. Second, the overvoltage itself may be forcibly held to an acceptable level by means of installation of surge arresters at the specific location to be protected. Third, the transmission system may be changed to minimize the effect of switching operations.

2. *Operation of circuit breakers.*[2] The maximum transient overvoltage that can reach the work site is often due to switching on the line on which work is being performed. If the automatic-reclosing is removed during energized line work so that the line will not be reenergized after being opened for any reason, the maximum switching surge overvoltage is then limited to the larger of the opening surge or the greatest possible fault-generated surge, provided that the devices (for example, insertion resistors) are operable and will function to limit the transient overvoltage. It is essential that the operating ability of such devices be assured when they are employed to limit the overvoltage level. If it is prudent not to remove the reclosing feature (because of system operating conditions), other methods of controlling the switching surge level may be necessary.

Transient surges on an adjacent line, particularly for double circuit construction, may cause a significant overvoltage on the line on which work is being performed. The coupling to adjacent lines must be accounted for when minimum approach distances are calculated based on the maximum transient overvoltage.

3. *Surge arresters.* The use of modern surge arresters has permitted a reduction in the basic impulse-insulation levels of much transmission system equipment. The primary function of early arresters was to protect the system insulation from the effects of lightning. Modern arresters not only dissipate lightning-caused transients, but may also control many other system transients that may be caused by switching or faults.

It is possible to use properly designed arresters to control transient overvoltages along a transmission line and thereby reduce the requisite length of the insulator string. On the other hand, if the installation of arresters has not been used to reduce the length of the insulator string, it may be used to reduce the minimum approach distance instead.[3]

4. *Switching restrictions.* Another form of overvoltage control is the establishment of switching restrictions, under which breakers are not permitted to be operated until certain system conditions are satisfied. Restriction of switching is achieved by the use of a tagging system,

---

[2]The detailed design of a circuit interrupter, such as the design of the contacts, of resistor insertion, and of breaker timing control, are beyond the scope of this appendix. These features are routinely provided as part of the design for the system. Only features that can limit the maximum switching transient overvoltage on a system are discussed in this appendix.

[3]Surge arrestor application is beyond the scope of this appendix. However, if the arrestor is installed near the work site, the application would be similar to protective gaps as discussed in paragraph IV.D. of this appendix.

similar to that used for a "permit," except that the common term used for this activity is a "hold-off" or "restriction." These terms are used to indicate that operation is not prevented, but only modified during the live-work activity.

D. Minimum approach distance based on control of voltage stress (overvoltages) at the work site. Reduced minimum approach distances can be calculated as follows:

1. *First method.* Determining the reduced minimum approach distance from a given withstand voltage.[4]

Step 1. Select the appropriate withstand voltage for the protective gap based on system requirements and an acceptable probability of actual gap flashover.

Step 2. Determine a gap distance that provides a withstand voltage[5] greater than or equal to the one selected in the first step.[6]

Step 3. Using 110 percent of the gap's critical flashover voltage, determine which electrical component of the minimum approach distance from Equation (2) or Table 6 which is a tabulation of distance versus withstand voltage based on Equation (2).

Step 4. Add the 1-foot ergonomic component to obtain the total minimum approach distance to be maintained by the employee.

2. *Second method.* Determining the necessary protective gap length from a desired (reduced) minimum approach distance.

Step 1. Determine the desired minimum approach distance for the employee. Subtract the 1-foot ergonomic component of the minimum approach distance.

Step 2. Using this distance, calculate the air gap withstand voltage from Equation (2). Alternatively, find the voltage corresponding to the distance in Table 6.[7]

---

[4]Since a given rod gap of a given configuration corresponds to a certain withstand voltage, this method can also be used to determine the minimum approach distance for a known gap.

[5]The withstand voltage for the gap is equal to 85 percent of its critical flashover voltage.

[6]Switch steps 1 and 2 if the length of the protective gap is known. The withstand voltage must then be checked to ensure that it provides an acceptable probability of gap flashover. In general, it should be at least 1.25 times the maximum crest operating voltage.

[7]Since the value of the saturation factor, a, in Equation (2) is dependent on the maximum voltage, several iterative computations may be necessary to determine the correct withstand voltage using the equation. A graph of withstand voltage versus distance is given in ANSI/IEEE Std. 516, 1987. This graph could also be used to determine the appropriate withstand voltage for the minimum approach distance involved.

TABLE 6   Withstand Distance for Transient Overvoltages

Crest voltage (kV)	Withstand distance (in feet) air gap
100...............	0.71
150...............	1.06
200...............	1.41
250...............	1.77
300...............	2.12
350...............	2.47
400...............	2.83
450...............	3.18
500...............	3.54
550...............	3.89
600...............	4.24
650...............	4.60
700...............	5.17
750...............	5.73
800...............	6.31
850...............	6.91
900...............	7.57
950...............	8.23
1000................	8.94
1050................	9.65
1100................	10.42
1150................	11.18
1200................	12.05
1250................	12.90
1300................	13.79
1350................	14.70
1400................	15.64
1450................	16.61
1500................	17.61
1550................	18.63

SOURCE: Calculations are based on Equation (2).
NOTE: The air gap is based on the 60-Hz rod-gap withstand distance.

Step 3. Select a protective gap distance corresponding to a critical flashover voltage that, when multiplied by 110 percent, is less than or equal to the withstand voltage from Step 2.

Step 4. Calculate the withstand voltage of the protective gap (85 percent of the critical flashover voltage) to ensure that it provides an acceptable risk of flashover during the time the gap is installed.

3. *Sample protective gap calculations.*

Problem 1:   Work is to be performed on a 500-kV transmission line that is subject to transient overvoltages of 2.4 pu. The maximum operating voltage of the line is 552 kV. Determine the length of the protective gap that will provide the minimum practical safe approach distance. Also, determine what that minimum approach distance is.

Step 1. Calculate the smallest practical maximum transient overvoltage (1.25 times the crest line-to-ground voltage):[8]

$$552 \text{ kV} \times \frac{\sqrt{2}}{\sqrt{3}} \times 1.25 = 563 \text{ kV}$$

This will be the withstand voltage of the protective gap.

Step 2. Using test data for a particular protective gap, select a gap that has a critical flashover voltage greater than or equal to

$$563 \text{ kV}/0.85 = 662 \text{ kV}$$

For example, if a protective gap with a 4.0-foot spacing tested to a critical flashover voltage of 665 kV (crest), select this gap spacing.

Step 3. This protective gap corresponds to a 110 percent of critical flashover voltage value of

$$665 \text{ kV} \times 1.10 = 732 \text{ kV}$$

This corresponds to the withstand voltage of the electrical component of the minimum approach distance.

Step 4. Using this voltage in Equation (2) results in an electrical component of the minimum approach distance of

$$D = (0.01 + 0.0006) \times \frac{552 \text{ kV}}{\sqrt{3}} = 5.5 \text{ ft}$$

Step 5. Add 1 foot to the distance calculated in step 4, resulting in a total minimum approach distance of 6.5 feet.

Problem 2: For a line operating at a maximum voltage of 552 kV subject to a maximum transient overvoltage of 2.4 pu, find a protective gap distance that will permit the use of a 9.0-foot minimum approach distance. (A minimum approach distance of 11 feet 3 inches is normally required.)

Step 1. The electrical component of the minimum approach distance is 8.0 feet (9.0–1.0).

Step 2. From Table 6, select the withstand voltage corresponding to a distance of 8.0 feet. By interpolation:

$$900 \text{ kV} + \left[50 \times \frac{(8.00 - 7.57)}{(8.23 - 7.57)} \right] = 933 \text{ kV}$$

---

[8]To eliminate unwanted flashovers due to minor system disturbances, it is desirable to have the crest withstand voltage no lower than 1.25 pu.

Step 3. The voltage calculated in Step 2 corresponds to 110 percent of the critical flashover voltage of the gap that should be employed. Using test data for a particular protective gap, select a gap that has a critical flashover voltage less than or equal to

$$933 \text{ kV}/1.10 = 848 \text{ kV}$$

For example, if a protective gap with a 5.8-foot spacing tested to a critical flashover voltage of 820 kV (crest), select this gap spacing.

Step 4. The withstand voltage of this protective gap would be

$$820 \text{ kV} \times 0.85 = 697 \text{ kV}$$

The maximum operating crest voltage would be

$$552 \text{ kV} \times \frac{\sqrt{2}}{\sqrt{3}} = 449 \text{ kV}$$

The crest withstand voltage of the protective gap in per unit is thus

$$697 \text{ kV} + 449 \text{ kV} = 1.55 \text{ pu}$$

If this is acceptable, the protective gap could be installed with a 5.8-foot spacing, and the minimum approach distance could then be reduced to 9.0 feet.

4. *Comments and variations.* The 1-foot ergonomic component of the minimum approach distance must be added to the electrical component of the minimum approach distance calculated under paragraph IV.D of this appendix. The calculations may be varied by starting with the protective gap distance or by starting with the minimum approach distance.

E. Location of protective gaps

1. Installation of the protective gap on a structure adjacent to the work site is an acceptable practice, as this does not significantly reduce the protection afforded by the gap.

2. Gaps installed at terminal stations of lines or circuits provide a given level of protection. The level may not, however, extend throughout the length of the line to the work site. The use of gaps at terminal stations must be studied in depth. The use of substation terminal gaps raises the possibility that separate surges could enter the line at opposite ends, each with low enough magnitude to pass the terminal gaps without flashover. When voltage surges are initiated simultaneously at each end of a line and travel toward each other, the total voltage on the line at the point where they meet is the arithmetic sum

of the two surges. A gap that is installed within 0.5 mile of the work site will protect against such intersecting waves. Engineering studies of a particular line or system may indicate that adequate protection can be provided by even more distant gaps.

3. If protective gaps are used at the work site, the work-site impulse insulation strength is established by the gap setting. Lightning strikes as much as 6 miles away from the work-site may cause a voltage surge greater than the insulation withstand voltage, and a gap flashover may occur. The flashover will not occur between the employee and the line, but across the protective gap instead.

4. There are two reasons to disable the automatic-reclosing feature of circuit-interrupting devices while employees are performing live-line maintenance:
   - To prevent the reenergizing of a circuit faulted by actions of a worker, which could possibly create a hazard or compound injuries or damage produced by the original fault:
   - To prevent any transient overvoltage caused by the switching surge that would occur if the circuit were reenergized.

However, due to system stability considerations, it may not always be feasible to disable the automatic-reclosing feature.

Appendix C to § 1910.269—Protection from Step and Touch Potentials

I. Introduction

When a ground fault occurs on a power line, voltage is impressed on the "grounded" object faulting the line. The voltage to which this object rises depends largely on the voltage on the line, on the impedance of the faulted conductor, and on the impedance to "true," or "absolute," ground represented by the object. If the object causing the fault represents a relatively large impedance, the voltage impressed on it is essentially the phase-to-ground system voltage. However, even faults to well grounded transmission towers or substation structures can result in hazardous voltages.[1] The degree of the hazard depends upon the magnitude of the fault current and the time of exposure.

---

[1]This appendix provides information primarily with respect to employee protection from contact between equipment being used and an energized power line. The information presented is also relevant to ground faults to transmission towers and substation structures; however, grounding systems for these structures should be designed to minimize the step and touch potentials involved.

II. Voltage-gradient distribution

A. Voltage-gradient distribution curve.   The dissipation of voltage from a grounding electrode (or from the grounded end of an energized grounded object) is called the ground potential gradient. Voltage drops associated with this dissipation of voltage are called ground potentials. Figure 1 is a typical voltage-gradient distribution curve (assuming a uniform soil texture). This graph shows that voltage decreases rapidly with increasing distance from the grounding electrode.

B. Step and touch potentials.   "Step potential" is the voltage between the feet of a person standing near an energized grounded object. It is equal to the difference in voltage, given by the voltage distribution curve, between two points at different distances from the "electrode." A person could be at risk of injury during a fault simply by standing near the grounding point.

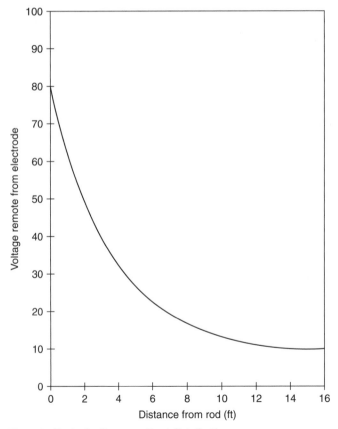

Figure 1   Typical voltage-gradient distribution curve.

"Touch potential" is the voltage between the energized object and the feet of a person in contact with the object. It is equal to the difference in voltage between the object (which is at a distance of 0 feet) and a point some distance away. It should be noted that the touch potential could be nearly the full voltage across the grounded object if that object is grounded at a point remote from the place where the person is in contact with it. For example, a crane that was grounded to the system neutral and that contacted an energized line would expose any person in contact with the crane or its uninsulated load line to a touch potential nearly equal to the full fault voltage.

Step and touch potentials are illustrated in Fig. 2.

C. Protection from the hazards of ground-potential gradients.   An engineering analysis of the power system under fault conditions can be used to determine whether or not hazardous step and touch voltages will develop. The result of this analysis can ascertain the need for protective measures and can guide the selection of appropriate precautions.

Several methods may be used to protect employees from hazardous ground-potential gradients, including equipotential zones, insulating equipment, and restricted work areas.

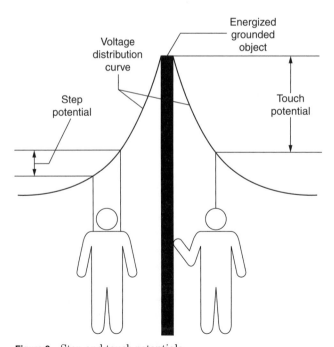

Figure 2   Step and touch potentials.

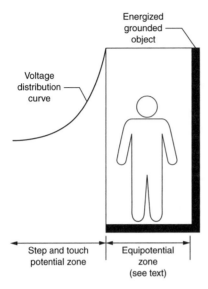

Energized
grounded
object

Voltage
distribution
curve

Step and touch
potential zone

Equipotential
zone
(see text)

Figure 3 Protection from ground-potential gradients.

1. The creation of an equipotential zone will protect a worker standing within it from hazardous step and touch potentials. (See Fig. 3.) Such a zone can be produced through the use of a metal mat connected to the grounded object. In some cases, a grounding grid can be used to equalize the voltage within the grid. Equipotential zones will not, however, protect employees who are either wholly or partially outside the protected area. Bonding conductive objects in the immediate work area can also be used to minimize the potential between the objects and between each object and ground. (Bonding an object outside the work area can increase the touch potential to that object in some cases, however.)

2. The use of insulating equipment, such as rubber gloves, can protect employees handling grounded equipment and conductors from hazardous touch potentials. The insulating equipment must be rated for the highest voltage that can be impressed on the grounded objects under fault conditions (rather than for the full system voltage).

3. Restricting employees from areas where hazardous step or touch potentials could arise can protect employees not directly involved in the operation being performed. Employees on the ground in the vicinity of transmission structures should be kept at a distance where step voltages would be insufficient to cause injury. Employees should not handle grounded conductors or equipment likely to become energized to hazardous voltages unless the employees are within an equipotential zone or are protected by insulating equipment.

Appendix D to § 1910.269—Methods of Inspecting and Testing Wood Poles

I. Introduction

When work is to be performed on a wood pole, it is important to determine the condition of the pole before it is climbed. The weight of the employee, the weight of equipment being installed, and other working stresses (such as the removal or retensioning of conductors) can lead to the failure of a defective pole or one that is not designed to handle the additional stresses.[1] For these reasons, it is essential that an inspection and test of the condition of a wood pole be performed before it is climbed.

If the pole is found to be unsafe to climb or to work from, it must be secured so that it does not fail while an employee is on it. The pole can be secured by a line truck boom, by ropes or guys, or by lashing a new pole alongside it. If a new one is lashed alongside the defective pole, work should be performed from the new one.

II. Inspection of wood poles

Wood poles should be inspected by a qualified employee for the following conditions:[2]

A. General condition.   The pole should be inspected for buckling at the ground line and for an unusual angle with respect to the ground. Buckling and odd angles may indicate that the pole has rotted or is broken.

B. Cracks.   The pole should be inspected for cracks. Horizontal cracks perpendicular to the grain of the wood may weaken the pole. Vertical ones, although not considered to be a sign of a defective pole, can pose a hazard to the climber, and the employee should keep his or her gaffs away from them while climbing.

C. Holes.   Hollow spots and woodpecker holes can reduce the strength of a wood pole.

D. Shell rot and decay.   Rotting and decay are cutout hazards and are possible indications of the age and internal condition of the pole.

---

[1] A properly guyed pole in good condition should, at a minimum, be able to handle the weight of an employee climbing it.

[2] The presence of any of these conditions is an indication that the pole may not be safe to climb or to work from. The employee performing the inspection must be qualified to make a determination as to whether or not it is safe to perform the work without taking additional precautions.

E. Knots.   One large knot or several smaller ones at the same height on the pole may be evidence of a weak point on the pole.

F. Depth of setting.   Evidence of the existence of a former ground line substantially above the existing ground level may be an indication that the pole is no longer buried to a sufficient extent.

G. Soil conditions.   Soft, wet, or loose soil may not support any changes of stress on the pole.

H. Burn marks.   Burning from transformer failures or conductor faults could damage the pole so that it cannot withstand mechanical stress changes.

III. Testing of wood poles

The following tests, which have been taken from § 1910.268(n)(3), are recognized as acceptable methods of testing wood poles:

A. Hammer test.   Rap the pole sharply with a hammer weighing about 3 pounds, starting near the ground line and continuing upwards circumferentially around the pole to a height of approximately 6 feet. The hammer will produce a clear sound and rebound sharply when striking sound wood. Decay pockets will be indicated by a dull sound or a less pronounced hammer rebound. Also, prod the pole as near the ground line as possible using a pole prod or a screwdriver with a blade at least 5 inches long. If substantial decay is encountered, the pole is considered unsafe.

B. Rocking test.   Applying a horizontal force to the pole and attempt to rock it back and forth in a direction perpendicular to the line. Caution must be exercised to avoid causing power lines to swing together. The force may be applied either by pushing with a pike pole or pulling with a rope. If the pole cracks during the test, it shall be considered unsafe.

Appendix E to § 1910.269—Reference Documents

The references contained in this appendix provide information that can be helpful in understanding and complying with the requirements contained in § 1910.269. The national consensus standards referenced in this appendix contain detailed specifications that employers may follow in complying with the more performance-oriented requirements of

OSHA's final rule. Except as specifically noted in § 1910.269, however, compliance with the national consensus standards is not a substitute for compliance with the provisions of the OSHA standard.

ANSI/SIA A92.2-1990, American National Standard for Vehicle-Mounted Elevating and Rotating Aerial Devices.

ANSI C2-1993, National Electrical Safety Code.

ANSI Z133.1-1988, American National Standard Safety Requirements for Pruning, Trimming, Repairing, Maintaining, and Removing Trees, and for Cutting Brush.

ANSI/ASME B20.1-1990, Safety Standard for Conveyors and Related Equipment.

ANSI/IEEE Std. 4-1978 (Fifth Printing), IEEE Standard Techniques for High-Voltage Testing.

ANSI/IEEE Std. 100-1988, IEEE Standard Dictionary of Electrical and Electronic Terms.

ANSI/IEEE Std. 516-1987, IEEE Guide for Maintenance Methods on Energized Power-Lines.

ANSI/IEEE Std. 935-1989, IEEE Guide on Terminology for Tools and Equipment to Be Used in Live Line Working.

ANSI/IEEE Std. 957-1987, IEEE Guide for Cleaning Insulators.

ANSI/IEEE Std. 978-1984 (R1991), IEEE Guide for In-Service Maintenance and Electrical Testing of Live-Line Tools.

ASTM D 120-87, Specification for Rubber Insulating Gloves.

ASTM D 149-92, Test Method for Dielectric Breakdown Voltage and Dielectric Strength of Solid Electrical Insulating Materials at Commercial Power Frequencies.

ASTM D 178-93, Specification for Rubber Insulating Matting.

ASTM D 1048-93, Specification for Rubber Insulating Blankets.

ASTM D 1049-93, Specification for Rubber Insulating Covers.

ASTM D 1050-90, Specification for Rubber Insulating Line Hose.

ASTM D 1051-87, Specification for Rubber Insulating Sleeves.

ASTM F 478-92, Specification for In-Service Care of Insulating Line Hose and Covers.

ASTM F 479-93, Specification for In-Service Care of Insulating Blankets.

ASTM F 496-93B, Specification for In-Service Care of Insulating Gloves and Sleeves.

ASTM F 711-89, Specification for Fiberglass-Reinforced Plastic (FRP) Rod and Tube Used in Live Line Tools.

ASTM F 712-88, Test Methods for Electrically Insulating Plastic Guard Equipment for Protection of Workers.

ASTM F 819-83a (1988), Definitions of Terms Relating to Electrical Protective Equipment for Workers.

ASTM F 855-90, Specifications for Temporary Grounding Systems to Be Used on De-Energized Electric Power Lines and Equipment.

ASTM F 887-91a, Specifications for Personal Climbing Equipment.

ASTM F 914-91, Test Method for Acoustic Emission for Insulated Aerial Personnel Devices.

ASTM F 968-93, Specification for Electrically Insulating Plastic Guard Equipment for Protection of Workers.

ASTM F 1116-88, Test Method for Determining Dielectric Strength of Overshoe Footwear.

ASTM F 1117-87, Specification for Dielectric Overshoe Footwear.

ASTM F 1236-89, Guide for Visual Inspection of Electrical Protective Rubber Products.

ASTM F 1505-94, Standard Specification for Insulated and Insulating Hand Tools.

ASTM F 1506-94, Standard Performance Specification for Textile Materials for Wearing Apparel for Use by Electrical Workers Exposed to Momentary Electric Arc and Related Thermal Hazards.

IEEE Std. 62-1978, IEEE Guide for Field Testing Power Apparatus Insulation.

IEEE Std. 524-1992, IEEE Guide to the Installation of Overhead Transmission Line Conductors.

IEEE Std. 1048-1990, IEEE Guide for Protective Grounding of Power Lines.

IEEE Std. 1067-1990, IEEE Guide for the In-Service Use, Care, Maintenance, and Testing of Conductive Clothing for Use on Voltages up to 765 kV AC.

[59 FR 4437, Jan. 31, 1994; 59 FR 33658, June 30, 1994; 59 FR 40729, Aug. 9, 1994]

NOTES

NOTES

Index